Films on
Solid Surfaces

Films on Solid Surfaces

The Physics and Chemistry of Physical Adsorption

J. G. DASH

Department of Physics
University of Washington
Seattle, Washington

ACADEMIC PRESS   New York   San Francisco   London   1975

A Subsidiary of Harcourt Brace Jovanovich, Publishers

ACADEMIC PRESS, INC.
111 Fifth Avenue, New York, New York 10003

United Kingdom Edition published by
ACADEMIC PRESS, INC. (LONDON) LTD.
24/28 Oval Road, London NW1

Library of Congress Cataloging in Publication Data

Dash, J G   (date)
    Films on solid surfaces.

    Bibliography: p.
    Includes index.
    1.   Adsorption.   2.   Thin films.   3.   Surface chem-
istry.   I.   Title.
QC182.D26   533'.1   74-30814
ISBN 0–12–203350–7

Contents

Preface

This book was written for the community of physical and biological scientists now working in fields involving physical adsorption and surface science. I have tried to cover most topics of contemporary interest, some by detailed treatment and others less extensively, but I have at least noted references to more complete accounts and guides to the current literature. The chapters are ordered so that the book might be used for graduate study following the plan of a course given at the University of Washington in 1973.

The subject is introduced by chapters on the atomic nature of physical adsorption and the states of single adsorbed atoms. There is a review of experimental methods for studying solid surfaces and films, and a discussion of substrate preparation. The equilibrium thermodynamics of surface films is given an extensive treatment and is developed from general statistical principles. Four chapters are given over to the various states of films and their phase transitions. The complications caused by substrate heterogeneity are discussed as a particularly challenging and important feature of all real films. Finally, there is a review of thin film superfluidity, with its still largely unanswered questions.

Acknowledgments

Many people helped to write this book, and I wish to indicate here my deep feelings of appreciation for all their contributions. The writing project was planned in collaboration with Frederick J. Milford. Although not able to continue due to the pressure of other commitments, he shares in the credit for its completion since I might never have begun it alone. I feel a great debt to my former students David L. Goodstein, G. Alec Stewart, Michael Bretz, and John A. Herb, who taught me more than I taught them. I have been fortunate to have as colleagues Michael Schick and Oscar E. Vilches, who with their independent talents in our common areas of interest have made our group a more exciting and rewarding place in which to work. It is a pleasure and an honor to thank Rudolf E. Peierls, who showed me how to think about films, and who encouraged me to write this book.

Permission to reprint various figures and tables has been generously granted by The American Chemical Society, The American Institute of Physics, Marcel Dekker, Inc., North-Holland Publishing Co., Taylor and Francis, Ltd., and John Wiley and Sons, Inc. I am grateful to L. L. Ban, J. K. Kjems, A. J. Melmed, and J. M. Thomas for making available photographic prints suitable for reproduction.

And finally, I thank my wife Joan, who has helped make my world more than two-dimensional.

1. Motivation

The study of thin surface films involves several fields of fundamental and practical importance—surface science, statistical thermodynamics, physics of condensed matter, and the experimental arts. Significant developments in each of these separate disciplines are now bringing about a revolution in film studies comparable to the advances in solid state physics of the 1930s. It is now possible to make detailed macroscopic and microscopic measurements of films on well-characterized solid surfaces, obtaining results that are highly reproducible from day to day in the same apparatus, and more significantly, in several different laboratories, and to correlate these results with the predictions of sophisticated theoretical models. Initial successes have encouraged more activity; fresh theoretical insights have stimulated new experiments, and reliable precision measurements have led to more realistic models. These advances have attracted new investigators into the study of films, bringing novel approaches and techniques, both experimental and theoretical, helping to generate an overall sense of enthusiasm and discovery.

What we find in the films is a rich succession of surface phases, some of which resemble two-dimensional gases, liquids, and solids, together with others that have no analogues in ordinary bulk matter. All are important proving grounds for theory. They provide physical examples of several theoretical models which were postulated many years ago but which until recently could not be subjected to experimental tests.

1

Much of the current interest has to do with the influence of dimensionality on long-range order. It has long been accepted that in 2D
matter there can be no perfectly ordered states or structures at any
finite temperature, which had been taken to imply that perfect 2D
crystals, magnets, superconductors, and superfluids cannot exist. These
sweeping implications have recently been called into question, and the
connections among various forms of long-range order and superfluidity,
magnetism, and crystallinity are now subjects for experimental and
theoretical investigation. Besides the question of long-range order, the
film phases can be used as novel systems to test ideas about conventional matter. Theoretical techniques previously developed for bulk
matter have recently been modified to 2D systems, and applied to
several examples of monolayer gases, liquids, and solids. These tests
have been strikingly successful in some cases and in others have suggested new paths toward the understanding of bulk matter. There also
are epitaxial phases in monolayers, in which the substrate structure
impresses some of its own regularity on the structure of the film. They
bear some resemblance to certain theoretical models which have been
studied intensively in connection with the phase transitions of bulk
matter. The Ising model of magnetism and the lattice gas model of
condensation occupy positions of great importance in the theory of
phase transitions in that exact solutions have been found for several
simple cases. Certain film phases show remarkable similarities to these
solutions in limited regimes but disagree in others, suggesting further
exploration of both theory and experiment.

The special regimes displayed by adsorbed films are not limited to
monolayer phases. Multilayers are in some ways more interesting and
complex than either the monolayer phases or bulk matter, for they
have the complications of both extremes. In multilayers there are the
interactions of bulk matter together with the strong gradients due to
substrate fields. In several systems there is evidence of distinctive
behavior of the individual layers of a two-layer and even three-layer
film. With greater substrate uniformity and lower temperatures it
should be possible to resolve the layers of even thicker films. These
results suggest some reformulation of the conventional thermodynamics of films and interfaces reaching down to fundamentals, even to
the operative definitions of "phase."

The increased resolution of modern studies brings about a heightened interest in substrate properties. Sophisticated theoretical models require increasingly detailed substrate characterization, and these theories are often subject to close comparison with experimental results. This new discrimination not only requires a far higher degree of substrate characterization than before, but permits physical adsorption to be developed into a more sensitive and searching probe of surface properties. Adsorption has been a most useful tool for studying surfaces, primarily used for gauging the effective areas and average attractive potentials of powders and porous bodies. But adsorption can yield far more information in favorable instances. Current models involve detailed knowledge of site symmetries and potentials, substrate phonon spectra, surface electronic densities, and adsorption heterogeneity. Heterogeneity can take many forms, involving variation in any of the substrate properties that affect the films. The spatial distribution of the heterogeneity can vary in magnitude, scale, and pattern. These variations could be described by a kind of topographic map of the surface, in which the altitude of the usual map is the substrate property under question. The topographies that characterize different surfaces reflect substrate treatment history and current condition, and the possible patterns can rival ordinary topographical maps in their variety of design and scale. Physical adsorption can in at least two thermodynamic regimes give some measure of these variations, perhaps eventually leading to a complete tracing out of the surface property.

The preceding description of modern physisorption focuses on fundamental questions without regard to their practical implications. However, we recognize that a certain degree of stimulation comes from the growing realization that surface science is extremely important to the needs of society. One can recognize several areas of application: adhesion, biological membranes, composite materials, heterogeneous catalysis, corrosion, fracture, lubrication, and thin film electronics. Each of these contains a strong element of surface science. They depend on the results of fundamental studies for their continued progress, they supply fresh problems and materials, and encourage the special efforts and satisfaction that can come with social relevance.

2. The Atomic Basis of Adsorption

2.1 THE ROLE OF THE SUBSTRATE IN PHYSISORPTION

Physical adsorption deals with the behavior of atoms on weakly attracting surfaces of bulk liquids and solids. The attraction is a general phenomenon, which is due to the fluctuating electric dipole moments mutually induced by all neutral materials. These interactions are in contrast to the stronger and more specific bonding involved in chemical adsorption. In physisorption the atoms and the surface are not strongly perturbed from their isolated states; this fact and the lack of specificity might suggest that the properties of physisorbed films can be understood without any detailed knowledge of the nature of the adsorbing surface. But this is not the case, for the interactions between the atoms and the substrate have to be judged relative to the interactions between the atoms in the film, and on this scale they are extremely important. Therefore the study of adsorbed films rests upon the understanding of the properties of the bare substrates and of their interactions with individual atoms.

The characterization of surfaces has advanced greatly within the past two or three decades, largely through the introduction and widening application of new experimental techniques, as indicated in Chapter 3. With these probes it is possible to determine, in principle at

5

least, virtually all of the properties of an experimental solid surface: the structure of the topmost layers, the amplitudes of the atomic vibrations, its perfection with respect to chemical impurities and crystalline order, and a good deal about the electronic structure.

Much of the new information comes as no surprise. The fact that the surface atomic structure is related to that of the bulk, that conduction electron distributions extend beyond the ion cores, and that most real surfaces are chemically and crystallographically imperfect are easily accepted. The main impact is in the depth and details of understanding of specific surfaces and gas–surface combinations. In many cases it is no longer necessary to invent a theoretical model for a substrate–gas system, since it is now more likely that a similar system, explored by means of one of the new microscopic techniques, is known to have a definite set of physical parameters. Therefore, as more microscopic results are obtained there will be greater interest in quantitative models, where substrate parameters can be introduced at the outset of the theoretical study of films. Nevertheless, current models of surfaces, including the most elementary, will continue to be useful. This is because there are certain domains of interest in which finer details of the substrate are relatively unimportant. For example, where the primary focus is on film properties that are largely determined by the interactions of the atoms when they are in a restricted geometry, it may be an unnecessary complication to include the periodic potential of the adatom–surface interactions. But with a developing appreciation for the properties of real surfaces, it will become increasingly important to justify the application of any simplified model to an actual experimental system.

The more general characteristics of substrates are contained within the several idealizations listed below. These categories are given roughly in order of increasing complexity; as mentioned earlier, each is still useful in modern theory. Detailed considerations bearing on the various models are taken up in subsequent sections of this chapter.

Models of solid surfaces

(i) *Plane boundary* A smooth, inert, mathematical surface; the simplest model of a real substrate. Although this idealization is not adequate in studies of very thin films, it can be a reasonable approximation in theories of surface effects on bulk phases, e.g., questions of restricted geometry.

(ii) *Attracting plane*   The *sine qua non* of physical adsorption is a surface-normal attraction between substrate and adsorbate atoms. This category contains two important subgroups.

(a) *Adhesion*   The assumption of adhesion without explicit account of the magnitude and range of the attraction forms the basis for the 2D abstraction of monolayer films. In this model one ignores surface-normal excitation of the film and of the vapor phase. For thicker films adhesion is assumed but gradients in the film are neglected; in such "slab" models the surface acts only as a boundary condition on the film.

(b) *Explicit attractive forces*   More realistic models of surface-normal forces are necessary when considering film–vapor equilibrium and multilayer formation. Variations of film properties with distance from the surface may become important in real systems, but are neglected in the lowest order approximation. When the attractive interaction is relatively strong in comparison to thermal energies this model approaches the 2D abstraction for monolayers.

(iii) *Adsorption sites*   As a first approximation to a structured surface, the substrate is treated as an array of potential wells or adsorption sites, each site capable of capturing a single atom. The internal energy structure of the sites may be ignored and the potential energy of the adatom taken to be a definite value with respect to the gas. This category forms the basis for the Langmuir theory of adsorption and for theories of 2D lattice gases and their ordering transitions.

(iv) *Structured substrate*   The atomic texture of the surface is treated more realistically than in (iii) by assuming a periodic variation of attractive energy along the two directions parallel to the surface. This model is the starting point for theories of localized-mobile transitions in films and for translational surface band states of adatoms.

(v) *Deformable substrate*   Models (i)–(iv) assume an inert and rigid substrate. Each of those categories can be made more realistic by the inclusion of excitations and deformations due to adsorption. Substrate excitations cause direct effects such as a heat capacity contribution and deformations modify the effective properties of the adsorbed atoms.

(vi) *Heterogeneity* All of the previous idealizations assume that the substrate is uniform along surface directions (or, as in (iii) and (iv), is periodic according to the assumed structure). Each category can be made more realistic by superimposing a certain degree of lateral variation along the surface.

2.2 GAS–SURFACE INTERACTIONS

All substrate categories beyond (i) assume an attraction between the gas and the surface, but the nature and form of the force law is often irrelevant to the problem under consideration. For monolayer properties it is often only necessary to consider an adhesion of a definite strength, and for certain problems in which substrate structure is important it may even be satisfactory to imagine a periodic potential without detailed attention to the origin of the periodicity. But for other properties, such as multilayer formation, the detailed form and magnitude of the interaction is central to the problem. Here we can separate the types of interactions into groups according to the nature of the surface.

2.2.1 van der Waals Solids

Low energy interactions between nonpolar neutral molecules are of the so-called London dispersion force or van der Waals type (1–4). At distances greater than about ten times the molecular diameter σ the interaction between pairs of molecules is attractive, the leading term varying as the inverse sixth power of the distance. The interaction arises from the attraction between the electric dipoles mutually induced by the fluctuating charge distribution of the atoms. The interaction strength is, to first approximation, proportional to the product of the atomic polarizabilities. At longer ranges, greater than $\sim$100 σ, retardation effects due to the finite transmission time of the fluctuation signals cause the interaction to decrease more rapidly, tending toward an inverse seventh power dependence. In the near range, where overlap of the electron clouds begins to be important, the interaction is specific for the types of atoms and their atomic states. Unfortunately it is this region that is of greatest importance to adsorption. The only general statement that can be made is that the potential has a minimum at a separation on the order of 1 Å and then rises at smaller distances. At still closer separations, in the region of strong overlap, the repulsion increases very rapidly. Theoretical calculations and semi-empirical fits to gas virial data for the close region of strong overlap in noble gases are approximated by either simple exponential or inverse power laws.

Although there are no general expressions valid over the entire range, a number of simple analytic expressions approximate the real potentials. One of the most traditional forms, useful for quantitative calculations, is the 6-12 Lennard-Jones potential

$$u(r) = 4\epsilon[(\sigma/r)^{12} - (\sigma/r)^{6}]. \qquad (2.2.1)$$

Dispersion forces are usually assumed to be pairwise additive, so that the interaction between a neutral nonpolar gas atom and the surface of a van der Waals solid is taken to be a simple sum of pair interactions

$$u(\mathbf{r}) = \sum_{i} u_{i}(|\mathbf{r} - \mathbf{r}_{i}|), \qquad (2.2.2)$$

where $\mathbf{r}_{i}$ represents the locations of the atoms of the solid. A simple approximation, valid for distances of several angstroms or more from the surface, is to treat the solid as a uniform continuum. On this basis the pair interaction Eq. (2.2.1) yields, for a plane semi-infinite solid and a gas atom at distance d above its surface,

$$u(d) = 4\pi\epsilon n_{s}[(\sigma^{12}/45d^{9}) - (\sigma^{6}/6d^{3}], \qquad (2.2.3)$$

where n_{s} is the atomic density of the solid. A potential of this form is illustrated in Fig. 2.1.

For distances smaller than a few lattice spacings the atomic structure of the substrate begins to be felt and the continuum approximation becomes inadequate. Several calculations for the near range have been made by explicit summation of pair energies. Most of the

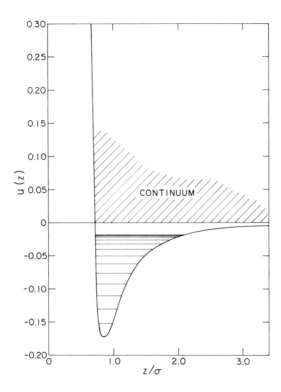

FIG. 2.1   Energy levels of an atom in the 9-3 potential due to the surface of an unstructured van der Waals medium. The large number of bound levels in this illustration would be characteristic of a strongly bound massive adsorbed atom.

computations are for noble gas atoms in "contact" with noble gas solids, assuming 6-12 Lennard-Jones interactions with parameters based on gas virial data (see Table 2.1). Values for the potential well depths $u_m(\sigma)$ typically have characteristic temperatures $u_m(\sigma)/k_B$ of several hundred degrees Kelvin, in rough agreement with estimates based on continuum approximations. The special characteristic of the summations is that the depth of the potential varies as the gas atom is moved across the surface, shown schematically in Fig. 2.2. This modulation is strongly dependent on the relative sizes of the atoms and the atomic density of the crystal face, being greatest for relatively small gas atoms on low density faces. As example, the variation is over a factor

Table 2.1 Parameters for 6–12 Pair Potentials of Simple Molecules[a]

Gas	σ (Å)	ϵ/k_B (°K)
He	2.556	10.22
Ne	2.789	35.7
Ar	3.418	124.0
Kr	3.61	190.0
Xe	4.055	229.0
N_2	3.681	91.5
CO	3.590	110.0
O_2	3.433	113.0
CO_2	3.996	190.0

[a] As tabulated by Hirschfelder et al. (2). Numerical values deduced from fits to experimental viscosity coefficients for all gases except He; for He the parameters are deduced from gas virial data.

two for a gas atom/solid atom size ratio of 0.7 on the (100) face of a face-centered cubic (fcc) crystal (5). The relative modulation amplitude is decreased for larger gas atoms and for denser surface planes, an effect which is illustrated in Table 2.2. The smoothing effect of denser surfaces is strong in the case of the basal plane of graphite. The nearest-neighbor spacing in the hexagonal net is only 1.42 A, about one-half of the argon atomic diameter; argon in contact with this surface is calculated to experience potential energy variations of only about 7% (5).

The dependence of the (x, y) periodic potential on the normal distance z means that the complete three-dimensional potential $u(x, y, z)$ is not separable into a sum of surface-parallel and surface-normal terms. Nevertheless, the assumption of separability allows one to explore simplified and instructive models, such as the 2D abstractions discussed in Chapter 5. In an early calculation of the atomic states of He adsorbed on a crystal surface, Lennard-Jones and Devonshire (6) assumed an analytic form for the 3D potential:

$$u(r) = D[e^{-2\kappa(z-\sigma)} - 2e^{-\kappa(z-\sigma)}] - 2\beta D e^{-2\kappa(z-\sigma)}[\cos ax + \cos ay], \quad (2.2.4)$$

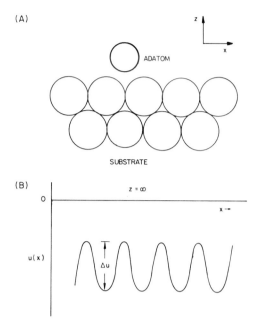

FIG. 2.2   Schematic illustration of the effect of substrate structure on the atom-surface interaction. As the adatom in (A) moves along the surface it experiences a varying potential energy (B). The modulation amplitude ΔV is strongly dependent on z: it may be comparable to the total binding for relatively small atoms in "contact" with the surface, but dies away to very small values at distances of a few atomic spacings.

which illustrates the nonseparability of surface-normal and parallel terms. Here D and κ are parameters describing the depth and fall-off of the z-wise potential, β is a small constant, and $2\pi/a$ is the lattice spacing.

2.2.2   Metals and Other Solids in the Continuum Approximation

The assumption of additivity of pair interactions, basic to the treatments of van der Waals gases and solids discussed in the previous section, is totally inadequate in the case of metals. In typical metals the conduction electrons form a highly correlated collective system which

Table 2.2 Variations of Binding
Energy of an Adatom
Moving along the
Surface of a Crystalline
Substrate[a]

| | Crystal face (%) | |
$\sigma(\text{gas})/\sigma(\text{solid})$	(100)	(111)
1.00	50	30
1.40	25	13

[a] Values tabulated refer to the extreme percentage variations in binding energy of noble gas atoms on faces of face-centered cubic rare gas solids (5). The values illustrate the smoothing effects of larger gas atoms and of denser surface planes.

cannot be considered as a superposition of independent particle contributions, even to a first approximation. Leaving aside the question of the ionic contribution momentarily, the interaction between a gas molecule and a metal surface might first be considered as if between the molecule and a continuous conducting half-space. This approach was taken by Lennard-Jones (7), who calculated the interactions between the fluctuating dipole moment of a neutral molecule and their images induced within the surface, finding that the potential varies as the inverse cube of the distance to the surface. Bardeen (8), in a detailed quantum treatment, also obtained the d^{-3} law but with an appreciably smaller magnitude, having shown that the earlier theory had neglected collective changes in the kinetic energies of all of the electrons near the external molecule. Prozen and Sachs (9) gave a quantum mechanical approximation suitable to semiconductors, where the electrons are nondegenerate, and they obtained an interaction energy varying as $d^{-2} \ln d$ for large separations. Casimir and Polder (10) showed that for perfectly conducting plane surfaces the effect of retardation is to change the interaction energy from d^{-3} at close distances to d^{-4} as $d \to \infty$.

All of these treatments approximate the metal as a continuum with particular electromagnetic properties. Lifshitz (11) pointed out that the continuum model should also be appropriate for other solids as well as metals, at separations sufficiently great to neglect the atomic structure of the material. The original theory has been extended as a more reliable basis for the calculation of interactions with any type of solid including van der Waals crystals, where pairwise additivity had been conventionally assumed (12–14). Lifshitz noted that in condensed bodies the close packing of the atoms materially changes the properties of their electronic envelopes, and the presence of some medium between the interacting atoms alters the electromagnetic field through which the interaction is effected. Instead of the conventional pairwise summation a macroscopic view is taken, in which the bodies are considered as continuous media; the solid is treated as a single, giant molecule. The basic idea of the macroscopic theory is still based upon an interaction via a fluctuating electromagnetic field, but the dielectric properties are taken to be those of the condensed bodies rather than a simple sum of free atom properties. The resulting interaction between two macroscopic bodies is dependent on their complex dielectric permeabilities. A general theory is obtained for the van der Waals forces, applicable to any bodies at any temperature, independent of their molecular nature (ionic or molecular crystals, amorphous bodies or liquids, metals, dielectrics, etc.), and it automatically includes retardation effects. The universal result is that the interaction between a neutral atom and a solid surface is as d^{-3} at moderate separations, changing to the retarded law d^{-4} at long range. The coefficients for the interaction energy are completely determined by the electromagnetic response functions of the molecule and the solid medium. The interaction energy $u(d)$ is expressed in terms of the electric polarizability α of the atom and the dielectric constant p of the body at imaginary frequencies $\omega = i\xi$. In the intermediate range where retardation is unimportant (14),

$$u(d) = -\frac{\hbar}{4\pi d^3} \int_0^\infty \alpha(i\xi) \left[\frac{p(i\xi) - 1}{p(i\xi) + 1} \right] d\xi. \qquad (2.2.5)$$

Quantitative tests of the complete theory, for He films adsorbed on crystals having well-characterized dielectric response functions, have

yielded excellent agreement with the measured adsorption energies
(15).

2.2.3 Metal Surfaces at Short Range

The interaction between a van der Waals gas molecule and a metal
surface presents great difficulties at short range, where fine scale features in the structure of the solid can be perceived in the external
potential. Such details are obviously beyond the scope of Lifshitz's
theory, which assumes the solid to be a homogeneous continuum.

In a real metal there is in addition to the conduction electron system
a lattice of ions, and the ionic contribution modulates the potential of
a nearby gas molecule with respect to translations parallel to the surface. This modulation might be estimated by the summation of pair
interactions, although the criticisms of pairwise additivity that are
made in the case of van der Waals solids are probably even more serious
in the present instance. The conduction electron distribution is likewise
modulated by the ionic lattice, so that a fraction of the periodic
potential is contributed by the conduction electrons and therefore the
entire potential is a collective property of the combined system. The
modulation of the electronic potential is generally smoother than that
due to the ions, since the electrons tend to "fill in" crevices between
the cores (16). Alternatively, if one describes the electron density by
a Fourier sum over all occupied electron states, the maximum wavevector is the k_F at the top of the Fermi distribution, which has a
magnitude on the order of an inverse lattice spacing in a typical metal.
This places an upper limit to the sharpness of the spatial modulation.

In addition to the modulations parallel to the surface the electronic
density has a certain variation normal to the surface, decreasing from
its average in the bulk to exponentially small values outside a distance
of a few angstroms, as shown in Fig. 2.3. This means that the conduction electron "surface" is diffuse, and furthermore, that it may not be
centered about the same position as the layer of ions. These last questions, of the diffuseness and location of the electronic distribution relative to the ionic lattice, are topics of current theoretical interest (17,
18). In traditional calculations the ions are usually treated as a uniform
positive background, i.e., the "jellium" model of a metal, but recent

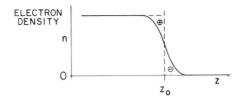

FIG. 2.3   The electron density in the neighborhood of a metal surface, according to the jellium model (17). z_0 marks the surface of the uniform positive charge background. The diffuseness and shift of the electron surface cause the electrostatic dipole layer characteristic of all metal surfaces.

calculations include an attempt to correct for the effect of the ion cores. For some purposes, e.g., for the calculation of the work function and intermediate range properties of the surface the jellium model appears to be a satisfactory approximation, but it is not at all an adequate basis for a description of the van der Waals potential at close range, where the lateral periodicity becomes important. Now it can be argued that the smoothing effect of the electrons is even more important to adsorbed molecules "in contact" with the surface than those at slightly greater distances. Here we are supposing that the gas molecule is a noble gas atom or similar molecule having a low polarizability. There is a repulsive interaction tending to prevent any overlap between the conduction electron density extending into the vacuum and the closed shell structure of the molecule. This repulsion, which arises from a combination of electrostatic interaction and the Pauli exclusion principle, may be strong enough to make the gas atom "float" on the conduction electron continuum. Since the conduction electron density is greater in the interstices between the ions, the total potential seen by the adsorbed atom is more uniform than that due to the ions alone. Thus, the surfaces of dense planes of metal crystals should be extremely smooth, both because of the fineness of the lattice structure and also because of the smoothing effect of the electrons. However, if the gas molecule is not simply a rigid closed shell structure, the interaction with the metal surface may well be much more specific as to both normal and lateral position and to the parameters of the electronic distribution. Such specificity is typical of chemisorption. As noted in the next chapter there are recent observations indicating that a great deal of specificity can sometimes occur even with noble gas atoms.

2.3 ADATOM STATES

2.3.1 Smooth Substrate Approximation

The energy level structure for surface-normal modes of an atom in the 9-3 potential [Eq. (2.3)] is shown schematically in Fig. 2.1. The rapid decrease of the potential energy with increasing z causes the level spacing to decrease rapidly at higher energies, blending into the continuum above $u = 0$. The actual energy level structure is not expressible in terms of simple functions, but a number of approximations can serve over limited ranges. One approximation fits the repulsive part by a hard-wall potential and a finite region of the attraction by a linear variation. In this scheme the solutions of the 1D Schrödinger equation can be expressed in terms of Airy functions, which are linear combinations of Bessel functions (19). A simpler procedure, valid for cases in which the atomic vibrations have small amplitudes, is to fit the curvature at the bottom of the well with a quadratic function and thus obtain the energy levels of the adatoms as 1D harmonic oscillators. This procedure yields for the force constant κ_z and frequency ω_z of an adsorbed atom of mass m

$$\kappa_z = 12\pi(N/V)_s \epsilon \sigma^6 z_m^{-5} = 12.2|u_m|/\sigma^2, \qquad (2.3.1)$$

$$\omega_z = (\kappa_z/m)^{1/2} = (3.49/\sigma)(|u_m|/m)^{1/2}. \qquad (2.3.2)$$

where

$$z_m = (\tfrac{2}{5})^{1/6}\sigma. \qquad (2.3.3)$$

For the harmonic approximation to be valid, the vibration amplitude must be a small fraction of z_m. In explicit terms of the quantum mechanical ground state mean squared amplitude $\langle z^2 \rangle_0$ of a harmonic oscillator, the condition for validity is

$$\langle z^2 \rangle_0 = \pi\hbar/m\omega_0 \ll z_m^2, \qquad (2.3.4)$$

where z_m^2 is given by Eq. (3.3).

It is interesting to compare numerical values of $\langle z^2 \rangle_0^{1/2}/z_m$ for light and heavy gases on a simple solid such as Kr. For the interaction parameters of dissimilar atoms we use the combining laws (2, 5)

$$\sigma_{xy} = \tfrac{1}{2}(\sigma_x + \sigma_y), \qquad \epsilon_{xy} = (\epsilon_{xx}\epsilon_{yy})^{1/2}, \qquad (2.3.5)$$

although more complex rules give somewhat better agreement with measurement (20). The solid is assumed to be a homogeneous material of the same density as solid Kr, and we use the empirical 6-12 parameters for the pure gases. The results are

$$\langle z^2 \rangle_0^{1/2}/z_m = 0.46 \qquad \text{for } {}^4\text{He atoms,}$$

$$\langle z^2 \rangle_0^{1/2}/z_m = 0.17 \qquad \text{for Ar atoms.}$$

These values indicate that the harmonic approximation is only fair for heavy atoms in their lowest states and is poor for He. For accurate calculations one must therefore use numerical methods.

2.3.2 Structured Surface: Localized Adatoms

A class of systems which are simply modeled involves strong binding in all three directions, so that the motions are restricted to the regions near the potential minima. There we may approximate the potential as harmonic in all three directions, although the force constants will in general be different. The surface normal force constant and frequency are estimated in the previous section, and here we obtain the parallel directions. For a simple illustration, we assume that the potential has the form at fixed $z = z_0$;

$$u(z_0 ; x, y) = u_m + u_1 \sin q_1 x + u_2 \sin q_2 y, \qquad (2.3.6)$$

where $q_{1,2} = 2\pi/a_{1,2}$ are the reciprocal lattice vectors for a rectangular surface net. Harmonic force constants are fitted to the regions of the minima by setting

$$\kappa_x = \left(\frac{\partial^2 u}{\partial x^2}\right)_{x_m}, \qquad \kappa_y = \left(\frac{\partial^2 u}{\partial y^2}\right)_{y_m}, \qquad (2.3.7)$$

where the positions of the minima are given by

$$q_1 x_m = q_2 y_m = \pi/2 \pm n\pi.$$

These relations yield for $u(z_0 ; x, y)$

$$\kappa_x = q_1{}^2 u_1 = 4\pi^2 u_1/a_1{}^2, \qquad \kappa_y = q_2{}^2 u_2 = 4\pi^2 u_2/a_2{}^2. \qquad (2.3.8)$$

These relations are to be compared with the surface normal force constant, Eq. (2.3.1). The lateral force constants may be comparable to κ_z for some crystal faces and atomic size ratios. For a typical example, we consider Steele's results (5) (Table 2.2) for gas atoms on a (100) face of an fcc crystal, at equal radii of gas atom and solid atom. In this case the modulation is 50% with a repeat distance $2^{1/2}\sigma$. Then, according to the harmonic approximation the lateral force constants $\kappa_x = \kappa_y = 4.93|u_m|/\sigma^2$, about 40% of κ_z. The lateral frequency is 64% of ω_z, i.e., comparable to but appreciably lower than the vibration frequency in the normal direction.

For the rectangular net chosen in this example the surface–parallel motions can be separated into two orthogonal modes, but this does not hold for typical surfaces. On the (100) face, for example, the surface net has fourfold symmetry about a site center with an angle of $\pi/4$ between the directions of maximum and minimum force constants. The sites on a (111) face have threefold symmetry; the angle between maximum force constant is $\pi/6$. With these symmetries, illustrated in Fig. 2.4, the 2D potential cannot be approximated merely as a sum of two quadratic terms, for there is also a cross term of comparable magnitude. The xy term represents a strong anharmonicity which invalidates a simple scalar representation in normal modes. The proper quantum mechanical solution can be obtained by numerical methods, using the complete 3D potential. Such calculations have been made by several groups in recent years, for He atoms adsorbed on rare gas solids and on basal planes of graphite. In most of these calculations the principal focus is on the tunneling of adsorbed atoms between sites, which is the topic of the next section.

2.3.3 Surface Band Structure

The states of an atom bound in a single isolated adsorption site form a set of discrete energy levels. On an adsorbing surface composed of many identical sites the proper quantum states are not discrete excitations within individual sites, but are bands of states of the indi-

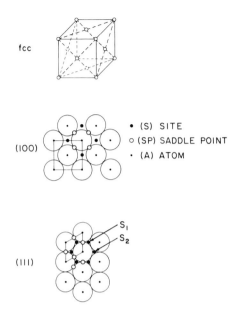

FIG. 2.4   Examples of site symmetries on three-dimensional crystal faces (5). The (111) face symmetry is more complicated than (100), with two classes of saddle points s_1 and s_2 which differ due to the underlying layer.

vidual atoms in the complete array of sites. The different states within each band represent tunneling translational states along the surface. Each band contains as many translational states as there are adsorption sites. The surface mobility of an adsorbed atom increases with the widths of the bands. These features are analogous to the general properties of conduction electrons in metals. In the following discussion we analyze a sequence of simple models borrowed from familiar examples and adapted to the problems of physisorption.

2.3.4   Lifetime in a Single Site

On a surface composed of identical sites the residence time of an atom adsorbed on a definite site is not arbitrarily long; it has a characteristic lifetime for escape to a nearby site, and will eventually make its way over all of the sites of the surface. This migration occurs even for strongly adsorbed atoms at zero temperature, by tunneling through the potential energy barriers between sites (21). An estimate of the characteristic tunneling time can be obtained by considering the problem of a particle in a simple 1D well, illustrated in Fig. 2.5A. Beginning from the strong-binding limit, the bound states of an atom of mass m have the kinetic energies of a particle in a box

$$K_n = [(\pi\hbar)^2/2md^2]n^2, \qquad n = 1, 2, \ldots . \qquad (2.3.9)$$

Potential barriers of finite width and height have a finite transmission coefficient: for deep states and strong barriers the coefficient is (22)

$$(16K_n/V_0^2)(V_0 - K_n)\exp(-G_n),$$

where

$$G_n = (2w/\hbar)[2m(V_0 - K_n)]^{1/2}. \qquad (2.3.10)$$

For barriers of arbitrary shape $u(x)$ the exponential factor is given approximately by (23)

$$G_n \simeq -(2/\hbar)\int_a^b [2m(u(x) - K_n)]^{1/2}\,dx, \qquad (2.3.11)$$

where the limits of integration are as shown in Fig. 2.5(B). The probability/unit time for escape from the well is the product of the transparency and the rate of hitting the barrier, hence the characteristic lifetime τ_n for tunneling out from the nth internal level is

$$\tau_n \approx t_n \exp(-G_n), \qquad (2.3.12)$$

where t_n is the transit time across the well. Thus the escape time is exponentially dependent on the mass of the particle and the "area"

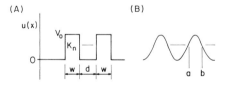

FIG. 2.5   Isolated adsorption sites. (A) Square potential barriers. (B) Site potential of more realistic shape. The barriers against escape of a particle from the bound level are the portions above the classical turning points a and b.

of the barrier above K_n; as is well known from nuclear physics this exponential sensitivity results in an enormous variation in lifetime for only moderate changes in the values of the parameters. For an illustration of this sensitivity in adsorption, the potential calculated by Ross and Steele (24) for He on (100) faces of solid Ar yields an estimated escape time from the lowest level of $\sim 10^{-6}$ in an "easy" direction between sites, whereas on the more closely packed (111) sites the time is five or six orders of magnitude smaller (21, 25). This decrease comes about from the combined effects of several factors: the decrease of barrier height and width, and also the narrowing of the well, which causes the transit time to decrease and the internal levels to move toward the tops of the barriers.

2.3.5 Two Sites; Level Splitting

Next we consider an atom in a double potential well representing two adjacent adsorption sites, as shown in Fig. 2.6. This example is familiar in textbook discussions of the one-dimensional double harmonic oscillator (26). The low-lying states n are each split into two levels, belonging to symmetric and antisymmetric wavefunctions ψ_{ns} and ψ_{na}. The symmetric state of each pair always has the lower energy. The energy difference between them is given by the overlap integral

$$\delta E_n = \int \left(|\psi_{na}|^2 - |\psi_{ns}|^2 \right) u(x) \, dx. \qquad (2.3.13)$$

Since the amplitudes within the barrier region increase with energy, the splitting is larger for states lying near the top of the central barrier. All of the states represent motion throughout the double well. A parti-

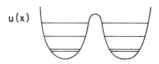

FIG. 2.6   Energy levels in a double well, showing the energy splitting between symmetric and antisymmetric states. The splitting energy δE_n is approximately equal to $\hbar/\tau_n$, where τ_n is the time for the particle to tunnel from one well to the other.

cle placed on one side eventually passes through the barrier to the
other side and after a time returns, producing a resonant interchange
of position between the two minima. The characteristic time for interchange of position is approximately equal to the escape time from a
single well having barriers of the same form as the central hump. This
time is related to the energy level difference δE_n by

$$\tau_n(\text{interchange}) \simeq \hbar/\delta E_n . \tag{2.3.14}$$

Equation (2.3.14) can be interpreted as an example of the uncertainty
principle, the energy uncertainty arising from the limited lifetime in a
single well. At temperatures much greater than $\delta E_n/k_B$ the two levels
of the nth state have essentially equal occupation probabilities. The
"splitting temperature" or "tunneling temperature" $\delta E_n/k_B$ is of
fundamental importance to the thermal properties of the system (see
Chapter 5). For numerical values of interest in adsorption, we note
that the splitting temperature corresponding to an interchange time
$\tau = 10^{-12}$ s is 7.6 °K; for 10^{-6} s the splitting temperature is 7.6 °μK.

2.3.6 Many Adsorption Sites; Bandwidths and Densities of States†

For a surface composed of N_s adsorption sites each level of the isolated site problem becomes split into N_s states of the many-site system. These states are grouped into bands, the energy width of each
band being comparable to the splitting between the two levels of the
double-well problem discussed previously. The stationary states of the
particle on the surface are not localized, but have equal probability
amplitudes on every identical site. The many-site problem is formally
identical to the situation of conduction electrons in a periodic lattice
except, of course, for the difference in dimensionality and quantitative
parameters. For adaptation of the theory of metals to adsorption we
have recourse to the standard theory (28), and we may use the results
of the various approximations, depending on the specific system of gas
and substrate. Some general features are illustrated here, emphasizing
those that are particularly important in adsorption.

† See Dash and Bretz (27).

In the tight-binding approximation one begins with the wavefunctions $\phi(r)$ of a particle in isolated potential wells and then constructs a wavefunction $\psi(r)$ for the single particle on the surface by a linear superposition

$$\psi = \sum_l c_l \phi(\mathbf{r} - \mathbf{r}_l). \tag{2.3.15}$$

The solutions of interest are of the Bloch form, i.e., the plane wave states

$$\psi_k = \sum_l c_{kl} \phi(\mathbf{r} - \mathbf{r}_l). \tag{2.3.16}$$

These satisfy the Bloch theorem if $c_{kl} = \exp(i\mathbf{k} \cdot \mathbf{r}_l)$, hence the unnormalized solutions are

$$\psi_k = \sum_l \exp(i\mathbf{k} \cdot \mathbf{r}_l) \phi(\mathbf{r} - \mathbf{r}_l). \tag{2.3.17}$$

The original hamiltonian H_0 is for isolated wells in which the particles are permanently localized. A perturbation term H' is now introduced, allowing for penetration. The perturbations cause each individual level of energy ϵ_0 to spread into a band of states, with individual energies

$$\epsilon_k = \epsilon_0 + \sum_l \exp(i\mathbf{k} \cdot \mathbf{r}_l) \int \phi^*(\mathbf{r}) H' \phi(\mathbf{r} - \mathbf{r}_l) \, d^3\mathbf{r}. \tag{2.3.18}$$

The only modification from the usual case of conduction electrons is that H' is independent of z and the integral reduces to one over x and y only; the states are plane waves in the two surface directions but have a definite energy with respect to z. The main features can be explored by limiting the summation to nearest neighbors only. Defining the integrals for the same site and a nearest-neighbor site as

$$\alpha \equiv \int \phi^*(\mathbf{r}) H' \phi(\mathbf{r}) \, d^2\mathbf{r}, \tag{2.3.19}$$

$$\gamma \equiv \int \phi^*(\mathbf{r}) H' \phi(\mathbf{r} - \mathbf{a}) \, d^2\mathbf{r}, \tag{2.3.20}$$

the energies can be evaluated for site arrays of definite symmetries. For a square array of side a, the nearest-neighbor atoms are at $(\pm a, 0)$

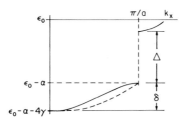

FIG. 2.7 Schematic illustration of the dispersion law $\epsilon(k)$ for a simple band structure, including the region of the first Brillouin zone boundary. The dotted line shows the quadratic law of the standard band approximation, which yields a constant effective mass over the entire first zone.

and $(0, \pm a)$: summing terms one obtains

$$\epsilon_k = \epsilon_0 - \alpha - 2\gamma(\cos k_x a + \cos k_y a). \qquad (2.3.21)$$

The dispersion relations $\epsilon(k_x)$ and $\epsilon(k_y)$ are shown schematically in Fig. 2.7 for the first Brillouin zone. The band structure is symmetric about $k = 0$. The next zone begins at $k = \pi/a$ with an upward energy displacement of Δ and has the same type of dependence of ϵ on k. Near the bottom and the top of each zone the curves can be approximated as quadratics. At the bottom of the first zone an expansion to second order yields

$$\epsilon_k = \epsilon_0 - \alpha - 4\gamma + \gamma a^2(k_x{}^2 + k_y{}^2). \qquad (2.3.22)$$

In this region the states of motion correspond to the two-dimensional translation of a free particle of effective mass

$$m^* = (\hbar k)^2/2\epsilon_k = \hbar^2/\gamma a^2. \qquad (2.3.23)$$

For an arbitrary point in the dispersion curve the effective mass is dependent on the direction and the curvature of the dispersion curve: it is a tensor quantity

$$m_{xy}{}^* = \hbar^2(\partial^2\epsilon/\partial k_x\,\partial k_y)^{-1}. \qquad (2.3.24)$$

Detailed calculations of thermal and transport properties require that the tensor m^* be carried through, but for less exacting purposes the effective mass may be approximated by a constant value $\bar{m}^*$ averaged over the entire band. The constant effective mass approximation makes the density of states $\overline{g(\epsilon)}$ uniform over the band: it corresponds to the density of states of a free particle of mass $\bar{m}^*$ in a field-free 2D space of area A,

$$\overline{g(\epsilon)} = A\bar{m}^*/2\pi\hbar^2, \qquad 0 \lesssim \epsilon \lesssim \delta, \qquad (2.3.25)$$

$$= 0, \qquad \epsilon > \delta.$$

Since the total number of states in the band is equal to the total number N_s of sites, we obtain two simple relations

$$\overline{g(\epsilon)} = N_s/\delta, \qquad \bar{m}^* = (2\pi\hbar^2/\delta)(N_s/A). \qquad (2.3.26)$$

Several quantitative calculations have been made of the He band structure adsorbed on various surfaces. The earliest was by Lennard-Jones and Devonshire (6), of ^{4}He on (100) planes of LiF. For input data they used molecular beam scattering results from which they estimated the z-wise well depth; from a previous lattice sum calculation the variation of potential along the surface was estimated at 30%. The band calculation was done in a weak-binding approximation. The approximation was justified by self-consistency; the particle motions were found to be nearly free in the surface directions, with an effective

Table 2.3   Calculated Bandwidths and Energy Gaps of ^{4}He Atoms Adsorbed on Various Solid Surfaces[a]

Surface	δ_1/k_B (°K)	δ_2/k_B (°K)	Δ_1/k_B (°K)	Reference
(100)LiF	~3	—	~3	6
(100)Kr	0.06	—	38.6	25
(100)Xe	~10^{-3}	—	42.9	25
(100)Kr	0.07	3.0	33.0	29
(100)Ar	0.1	3.0	38.0	29
(111)Ar	4.9	26.0	22	31
Graphite	~7	—	~0	32
Xe-plated graphite	~8	~20	~7	33
Ar-plated Cu	~15	—	~0	34

[a] δ_1, δ_2 refer to the lowest and first excited bandwidths, respectively, and Δ_1 is the energy gap between them.

mass at the bottom of the lowest band only 8% higher than the bare atomic mass.

Calculations of He band structure on rare gas solid surfaces have been carried out by several groups of investigators within recent years (25, 29–34), and all have found somewhat smaller mobilities than the LiF results, although in all cases the tunneling was found to be appreciable. All of the modern calculations are based upon pairwise summations of 6-12 power laws with empirical coefficients. In Table 2.3 we list the various model substrates which have been studied, with the widths of the lowest and first excited band and the first energy gap.

2.4 EFFECTS OF SUBSTRATE DEFORMATION

The properties of adsorbed atoms may differ from those in the gas phase due to indirect processes involving the substrate. Several mechanisms have been recognized but only one has been conclusively detected in an experiment. Nevertheless, order of magnitude estimates indicate that induced changes may be quite important in some circumstances; for further discussion of magnitudes and detectability, see Chapter 6.

2.4.1 Induced Static Dipole Moments; Changes of Work Function†

The work function of a metal surface is changed by the adsorption of small amounts of gas. The effect is quite large and is observed for all combinations of gases and metals. For the noble gases and other van der Waals molecules the change is always such as to reduce the work function and its magnitude is proportional to the quantity adsorbed at coverages much less than a monolayer. There is little if any change at greater coverage. This observation implies that there is an induced static dipole moment associated with each adsorbed atom in the first layer, oriented with positive ends away from the surface. But although the effect has been known for many years there is no satisfactory theory. Several mechanisms have been proposed, two of which

† See refs. (35–41).

were suggested by Mignolet (35): the formation of a virtual chemical bond between the atom and the surface, or the polarization of the neutral molecule by the electric field of the surface layer. Each of these explanations has modern proponents, but no definitive experimental tests have yet been put forth. Antoniewicz (41) has recently suggested a third mechanism: that the induced dipole occurs because the orbital electrons of the atom tend to be displaced toward the metal by the image charge. An estimate of magnitude is in rough agreement with observation. Still another possible mechanism involves perturbation of the dipole surface charge of the metal by the adsorbed atom (42). The dipole surface charge is produced by the spilling out into the vacuum of some of the electron density of the metal. A noble gas atom approaching the surface will first make contact with this leakage charge, and owing to the quantum mechanical repulsion with the closed shell configurations, the charge tends to be forced back into the metal.

Although the explanation is in dispute, the observations of large work function changes in physisorption carries important implications for both the understanding of metal surfaces and for the physics of films. As far as films are concerned, the static moments must cause a novel long-range repulsion between adatoms (43). At large separations they must appear as point dipoles, with interaction energy varying as r^{-3}. This interaction is much more gradual than the van der Waals attraction and hence will tend to dominate at very low coverages. At moderate and high coverage the dipole interaction can be much more complicated. The r^{-3} dependence will hold for separations large compared to the "size" of each dipole, i.e., the region of the disturbance in the dipole layer. The "size" of an induced dipole is controlled by the screening length of the surface. Any externally induced disturbance in the conduction electron distribution at the surface is screened out in a distance usually on the order of a few lattice spacings, although in certain systems the screening length may be quite large (44, 45). The characteristic response is generally an exponential variation at close distances, followed by slowly damped oscillations. Whatever the actual screening length and its detailed variation might be, this type of spatially extended response causes the induced static dipole moments of two atoms to have a different interaction law at moderate and short ranges, when their screened dipole moments merge. At separations of several screening lengths, where the damped oscillations overlap, the interaction energy should tend to oscillate with r, since it corresponds to merging of two polarization clouds of sometimes parallel and sometimes antiparallel orientations. These oscillations may be much smaller than the monotonic repulsions coming from the nonoverlapping portions. At distances equal to the screening length or less, the polarization clouds have the same polarity in the major region of overlap, and hence the interaction energy will increase monotonically but less rapidly than as r^{-3}. Finally, at very close distances, where the unshielded disturbance in the conduction electron density is seen by each atom, the interaction should tend toward an inverse first power law. Of course the above discussion of an extended "size" to the dipole is relevant only to the conduction electron contribution; for that part due to polarization of the atoms themselves, the interaction remains as r^{-3} all the way to "contact."

2.4.2 Correlations of Fluctuation Moments

The primary interaction between a surface and a neutral molecule is via the fluctuating electromagnetic fields; it is attractive at moderate and long range. Casimir and Polder (10) pointed out that in the neighborhood of a conducting surface the interaction between two van der Waals gas molecules is reduced by their image dipoles induced in the metal. Sinanoğlu and Pitzer (46) showed that there ought to be such an effect with any polarizable medium. As McLachlan (14) has proved, it is relatively simple to calculate using the general theory originated by Lifshitz. This "surface-mediated" or "indirect" interaction is a kind of high-frequency extension of the static dipole repulsion discussed in the previous section. In this case the fluctuating induced moments are less attractive whenever they are correlated in (retarded) time, i.e., in phase with each other.

The in-phase components contribute a repulsion to two atoms side-by-side on the surface, but an attraction to atoms aligned along z, i.e., when they are vertically above one another in adjacent layers. The interaction for two side-by-side atoms varies as r^{-6} at moderate and long range; for very short separations, actually smaller than atomic diameters would permit, the theory predicts a r^{-2} law. Magnitudes for typical materials indicate that the coefficient of the r^{-6} term might be on the order of 20% of the normal van der Waals attraction, hence the attraction would be some 20% lower for adatoms than for the same atoms in the gas phase. Experiments to detect such differences have been inconclusive; there is some discussion of attempts in Chapter 6.

2.4.3   Effective Mass

There are two distinct ways in which the apparent mass of an adsorbed atom is changed by the substrate. The first, already discussed in Section 2.3.3, is an effect of the band structure. [If we restrict the band model to cases of very low coverage, where interactions between the adatoms can be neglected (see Chapter 5), the band mass is on the average equal to or greater than the bare mass, although for certain regions of the band it may be smaller than m.] An entirely different mass change is due to the compressibility of the substrate, always causing an increase. Since an adsorbed atom "weighs down" upon the surface there is a local pressure which disturbs the positions of the atoms of the solid in the neighborhood of the adatom; there is a local depression or "dent" in the surface. This distortion is carried about wherever the adatom moves along the surface. Thus there is an associated motion of the atoms of the solid along with the adatom, and this adds to the effective mass of the adatom. One can speak of the distortion as due to a cloud of phonons localized near the adatoms. To distinguish this mass increase from the band mass we can describe it as a "phonon mass" contribution, and use m^* to denote the total, including both band and phonon changes. The phonon mass increase for adatoms is analogous to the increased effective mass of polarons in semiconductors and of electrons moving through an ionic lattice (47), although of course the primary forces causing the distortions are entirely different.

2.4.4   Substrate Phonon Exchange

The surface distortion which increases the effective mass of an adatom can interact with a similar distortion of another adatom. This interaction is generally attractive. It can be viewed as if each adatom tends to roll into each other's dent, and has been called the "mattress effect" (48). A more sober name is "substrate phonon exchange" (49). Theoretical estimates indicate that it is a very long-ranged interaction, varying approximately as the inverse square of the separation between the adatoms. For He on a model surface representing a typical adsorbent, the exchange interaction is comparable to the direct interaction at distances of several angstroms. Therefore, one might expect to see evidences of exchange at moderately low coverage, but none have been obtained as yet.

REFERENCES

1. A. Dalgarno and W. D. Davison, *Advan. At. Mol. Phys.*, **2**, 1 (1966).

2. J. O. Hirschfelder, C. F. Curtiss, and R. B. Bird, "Molecular Theory of Gases and Liquids." Wiley, New York, 1954.

3. J. O. Hirschfelder (ed.), "Intermolecular Forces." Wiley, New York, 1967.

4. I. A. Torrens, "Interatomic Potentials." Academic Press, New York, 1972.

5. W. A. Steele, *Surface Sci.* **34**, 41 (1972).

6. J. E. Lennard-Jones and A. F. Devonshire, *Proc. Roy. Soc. (London)* **A158**, 242 (1937).

7. J. E. Lennard-Jones, *Trans. Faraday Soc.* **38**, 333 (1932).

8. J. Bardeen, *Phys. Rev.* **58**, 727 (1940).

9. E. J. R. Prozen and R. G. Sachs, *Phys. Rev.* **61**, 70 (1942).

10. H. B. G. Casimir and D. Polder, *Phys. Rev.* **73**, 360 (1948).

11. E. M. Lifshitz, *J. Exp. Theoret. Phys. USSR* **29**, 94 (1955) [*English transl.: Sov. Phys.-JETP* **2**, 73 (1956)].

12. I. E. Dzialoshinskii, E. M. Lifshitz, and L. P. Pitaevskii, *J. Exp. Theoret. Phys.* **30**, 1152 (1956) [*English transl.: Sov. Phys.-JETP* **3**, 977 (1957)]; *Advan. Phys.* **10**, 165 (1961).

13. C. Mavroyannis, *Mol. Phys.* **6**, 593 (1963).

14. A. D. McLachlan, *Mol. Phys.* **7**, 381 (1964).

15. E. S. Sabisky and C. H. Anderson, *Phys. Rev.* **A7**, 790 (1973).

16. R. Smoluchowski, *Phys. Rev.* **60**, 661 (1941).

17. N. D. Lang and W. Kohn, *Phys. Rev.* **B1**, 4555 (1970); **B3**, 1215 (1971); **B7**, 3541 (1973).

18. V. Heine and C. H. Hodges, *J. Phys. C: Solid State Phys.* **5**, 225 (1972).

19. M. Abramowitz and I. A. Stegun, *"Handbook of Mathematical Functions,"* Nat. Bur. Std., Washington D.C., 1964; Dover, New York, 1965.

20. D. D. Konalow, *J. Chem. Phys.* **50,** 12 (1969).

21. J. G. Dash, *J. Chem. Phys.* **48,** 2820 (1968).

22. L. Schiff, "Quantum Mechanics." McGraw-Hill, New York, 1949.

23. E. Fermi, "Nuclear Physics," pp. 55–65. Univ. of Chicago Press, Chicago, Illinois, 1949.

24. M. Ross and W. A. Steele, *J. Chem. Phys.* **35,** 862 (1961).

25. F. Ricca, C. Pisani, and E. Garrone, *J. Chem. Phys.* **51,** 4079 (1969).

26. E. Merzbacher, "Quantum Mechanics," 2nd ed., pp. 65–78. Wiley, New York, 1970.

27. J. G. Dash and M. Bretz, *Phys. Rev.* **174,** 247 (1968).

28: See, for example, N. F. Mott and H. Jones, "The Theory of the Properties of Metals and Alloys." Reprint Edition. Dover, New York, 1958; or W. A. Harrison, "Solid State Theory." McGraw-Hill, New York, 1970.

29. A. D. Novaco and F. J. Milford, *J. Low Temp. Phys.* **3,** 307 (1970); see also H. Hollenbach and E. E. Salpeter, *J. Chem. Phys.* **53,** 79 (1970).

30. H. W. Lai, C. W. Woo, and F. Y. Wu, *J. Low Temp. Phys.* **3,** 463 (1970).

31. H. W. Lai, C. W. Woo, and F. Y. Wu, *J. Low Temp. Phys.* **5,** 499 (1971).

32. D. E. Hagen, A. D. Novaco, and F. J. Milford, "Adsorption–Desorption Phenomena" (F. Ricca, ed.). Academic Press, New York, 1972.

33. A. D. Novaco and F. J. Milford, *Phys. Rev.* **A5,** 783 (1972).

34. F. J. Milford and A. D. Novaco, *Phys. Rev.* **A4,** 1136 (1971).

35. J. C. P. Mignolet, *J. Chem. Phys.* **21,** 1298 (1953); *Discuss. Faraday Soc.* **8,** 105 (1950)

36. J. H. de Boer, *in* "Chemisorption" (W. E. Garner, ed.), p. 171. Butterworths. London and Washington, D.C., 1957.

37. P. M. Gundry and F. C. Tompkins, *Trans. Faraday Soc.* **56,** 846 (1960).

38. P. G. Hall, *Chem. Commun.* No. 23, 877 (1966).

39. G. Ehrlich and F. G. Hudda, *J. Chem. Phys.* **30,** 493 (1959).

40. T. Engel and R. Gomer, *J. Chem. Phys.* **52,** 5572 (1970).

41. P. R. Antoniewicz, *Phys. Rev. Lett.* **32,** 1424 (1974).

42. Unpublished notes of the Author.

43. J. Topping, *Proc. Roy. Soc. (London)* **A114,** 67 (1927).

44. P. B. Visscher and L. M. Falicov, *Phys. Rev.* **B3,** 2541 (1971).

45. E. Canel, M. P. Matthews, and R. K. P. Zia, *Phys. Kondens. Mater.* **15,** 191 (1972).

46. O. Sinanoğlu and K. S. Pitzer, *J. Chem. Phys.* **32,** 1279 (1960).

47. C. Kittel, "Quantum Theory of Solids," Chapter 7. Wiley, New York, 1963.

48. G. A. Stewart and J. G. Dash, *Phys. Rev.* **A2,** 918 (1970).

49. M. Schick and C. E. Campbell, *Phys. Rev.* **A2,** 1591 (1970).

3.   Experimental Techniques and Substrates

3.1   EXPERIMENTAL TECHNIQUES

The experimental study of adsorption has advanced considerably in recent years, largely through the introduction and widening application of several powerful techniques for characterizing the microscopic nature of substrates and films. Whereas the experimenter had once been limited to thermal and vapor pressure measurements, modern surface science includes low energy electron diffraction, field emission and ionization microscopy, Auger electron spectroscopy, and low energy molecular scattering. The list is long and is still expanding. It has been estimated that there are some thirty recognized techniques which involve the bombardment of a surface by molecules, ions, electrons, or photons, and the analysis of resulting emission products (1). Some of them have already yielded important results which had been only suspected from earlier studies, and in certain cases have shown older views to have been incorrect. The new techniques do not completely supplant older methods, however, for they are each subject to definite limitations, so that while each may offer a highly detailed view of the surface, it is generally only an aspect of the total picture. A complete description of a specific adsorption system and of adsorption in general requires the use of several types of surface probes among which both older and newer methods play essential roles. In the following paragraphs we give brief descriptions of the more widely applied or most promising of the techniques, with some of their principal advantages and limitations.

Of all of the modern techniques, the most widely used is low energy electron diffraction (LEED). There is an extensive bibliography stretching back to 1927, which shows a great increase in activity in the 1960s (2), when commercial instruments became available. LEED involves the bombardment of a surface by monoenergetic electrons with energies of a few hundred volts, and analysis of the scattered electrons (3). Owing to the low penetrating power of the electrons, LEED gives a view of only top layers of a surface. As an example, for 200-V electrons on Au, the first monolayer contributes about 50% of the total diffracted intensity and the first two monolayers together provide about 90%. The area of the diffracting region, which is primarily determined by the focussing of the electron beam, is typically ~ 1 mm^2, and the beam can be deflected to explore selected regions of a larger crystal surface. The symmetries and sharpness of a diffraction pattern can be interpreted to yield the structure of the surface. By studying the temperature dependence of the diffraction spots and the diffuse background, one can gain some insight into the dynamics of the surface atoms (4, 5). LEED is an extremely useful diagnostic tool in studies of surface preparation and of ordered structures of adsorbed gases. Difficulties include uncertainties in the interpretation of patterns (6) and desorption of weakly bound films by the electron beam (7). LEED requires high vacuum for operation and cannot be applied to systems having vapor pressures higher than about 10^{-5} Torr. It is not, therefore, generally suited to physisorption, although successful studies have been made of strongly bound molecular gases (8, 9). It is sensitive to concentrations larger than a few percent of a full monolayer. LEED has several variants which employ beam energies outside the conventional range, and these are sometimes given acronyms of their own: VLEED (very low); MEED (medium); HEED (high). RHEED, standing for "reflection high energy electron diffraction," uses both higher beam energies and grazing incidence: this technique can examine somewhat deeper layers of a substrate and provide a measure of longer range order than is possible with LEED.

Auger electron spectroscopy (AES) is based on the distinctive electron emission spectra of atoms returning to their ground states after inner shell excitation. The process was first pointed out by Lander (10) as potentially valuable for surface analysis since the escaping electrons originate within the top few layers. A succession of technical modifications and improvements by several individuals and groups have made it an extremely valuable probe of surface chemical composition (11, 12). The exciting electrons typically have an energy of 2000 eV and beam diameter of 0.1 mm. It is capable of very high sensitivity: in a recent study an adsorption of $\sim 1/500$ layer or 10^{10} atoms could be detected (13). Owing to the requirement of high vacuum and the tendency to cause desorption or excitation of weakly bound atoms, AES has limited utility for studies of physisorbed films. However it has been used very successfully in combination with LEED for a study of Xe adsorption at low temperature (13).

Electron spectroscopy for chemical analysis (ESCA) is similar to AES, but differs in that the excitation is by x radiation of discrete energy and both photo and Auger electrons are analyzed (14, 15). Both AES and ESCA can detect small energy shifts due to changes in valence state of the surface atoms. They are extremely useful in the study of chemical composition of bare and chemisorbed substrates. Ion neutralization spectroscopy (16, 17) is still more sensitive than AES to the chemical state of the surface atoms. In this method an ion beam is neutralized by electrons tunneling out of the surface; the tunneling probability is strongly dependent on the condition of the outer electronic shells.

Field emission microscopy (FEM) (18–21) is based upon the field extraction of electrons from a fine tip electrode (radius ~ 1000 Å) and their subsequent collection, energy analysis, or imaging on a phosphor screen. Variations in the barrier for electron tunneling out of the surface can cause the field emission current to vary in magnitude and energy. The spatial resolution is limited by quantum diffraction to about 10 Å, hence it cannot see detail as fine as the atomic structure. However, moderate-size adsorbed molecules cause diffraction patterns which reflect the symmetries of their principal structural components (Fig. 3.1). However, it is useful for examining differences between distinct crystal facets. It is also capable of detecting the presence of adsorbed layers, through the changes they cause in the work function of

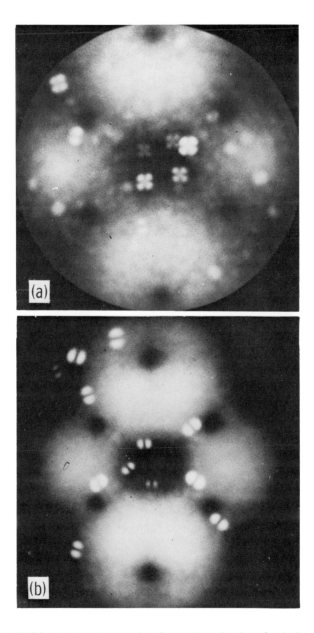

FIG. 3.1   Field emission micrographs of organic molecules adsorbed on tungsten at 77 K. (a) Copper phthalocyanine. (b) Flavanthrene (22). (Courtesy of A. J. Melmed, National Bureau of Standards.)

the surface and the tunneling barrier (22). The method is subject to the usual problems of ultrahigh vacuum measurements, and FEM cannot be used for films having vapor pressures greater than about 10^{-6} Torr. The high fields necessary for electron emission cause field desorption of weakly bound adatoms; but it is possible to obtain views of such films by intermittent operation, and measurements of binding energies can be made by applying fields large enough to cause appreciable desorption.

The highest spatial resolution, on the order of 3 Å, is possible in field ion microscopy (FIM) (18, 19). In this case a small amount of imaging gas, usually He, is introduced to the system. Gas atoms are ionized within a few angstroms of the surface by electrons tunneling into the surface, and the subsequent ion current is collected or displayed in a manner similar to FEM. FIM provides a direct visual picture of the atomic structure of the surface; images of individual atomic positions of the top surface layer. The distinctive arrangements of different crystal facets can be seen, with clear indications of any dislocations, vacancies, or adsorbed atoms (Fig. 3.2). With FIM one can detect and observe diffusion and desorption of individual atoms of the substrate or adsorbed film (23, 24). The detailed process of ionization is not well understood. Recently E. W. Muller, who invented both FEM and FIM, has added mass spectrometry to field ion microscopy (25); with this refinement known as "atom-probe FIM," it is possible to identify single ions field desorbed from the tip.

Low energy molecular scattering, although first applied to adsorption over 40 years ago (26, 27), is a relatively recent addition to the modern repertory (28, 29). Its capabilities, which are still in the process of development, are virtually unique in providing direct measurements of the atomic variations in potential between single neutral adatoms and solid surfaces. These variations, which are of controlling importance to the surface mobility of physisorbed atoms, cause diffraction peaks in the scattering of monoenergetic atomic beams. The technique has recently been successful in detecting the ordered structures of physisorbed films (30) as well as of a few bare substrates (31). But present limitations include severe problems of surface contamination (32) and the lack of a theoretical understanding of energy transfer mechanisms; the process is still a subject for study rather than an experimental probe of adsorption.

FIG. 3.2   He ion field emission photograph of a tungsten crystal surface (56).

Ellipsometry was suggested at the end of the last century by Drude
(33) as an outgrowth of his method for determining the optical constants of bulk materials. The technique, which is capable of measuring
the thickness and index of refraction of thin adsorbed films down to
monolayer coverages, involves changes in the state of polarization of
light upon reflection from the surface. It has been refined considerably
in recent years, and is gaining wider application in studies of adsorption (34), offering particular advantages when combined with other
techniques such as LEED and FEM (35).

Nuclear magnetic resonance and electron paramagnetic resonance
(36) have been widely applied to studies of bulk materials; although
these techniques should be similarly useful in surface studies such ap-

plications have been relatively few in number. Perhaps their most distinctive contribution to physical adsorption can be as indicators of adatom mobility, through measurement of line widths and relaxation times. An nmr study of ^{3}He on graphitized carbon black has given clear evidence of mobility changes (37, 38) (see Chapters 7 and 8), but studies of other gases and substrates have not produced comparably distinctive results (39, 40).

Mössbauer spectroscopy (41) offers some capabilities in common with nmr and esr, and also has unique potentialities for physical adsorption (42). For example, the amplitude of the recoilless fraction can be used to gauge the strength and harmonicity of the adatom–surface interaction. There are two isotopes which have particularly suitable characteristics for physical adsorption: ^{83}Kr and ^{129}Xe. A recent experiment was performed with the krypton isotope (43), and further studies are under way.

As the most recent addition to the list we take note of neutron diffraction of thin films. In a recent study of N_2 adsorbed on graphite several distinct ordered and disordered monolayer structures were observed (44, 45), and portions of the phase boundaries could be correlated with changes in the vapor pressure. Neutron diffraction opens up for study the structural arrangements of many physisorbed films that are not accessible to LEED because of high vapor pressures at the temperatures of interest or because the electron beam causes desorption. As a further benefit, inelastic neutron scattering might be used to determine the excitation spectra of the films; active research is now exploring this possibility.

In comparison with the newer techniques, traditional measurements of vapor pressure, heats of adsorption and heat capacity are coarse-grained probes, which yield no direct information on the microscopic states of the film or substrate. Interpretation of such measurements are therefore particularly dependent on theoretical models. Nevertheless, an adsorption system is never fully characterized unless its thermodynamic properties have been studied, by either vapor pressure or calorimetric measurements, or preferably, both. It is sometimes necessary and always advisable to supplement them with other investigations of the same systems. Another drawback is that these methods typically require much larger experimental adsorption areas than most of those previously discussed. Typical areas required for vapor pressure studies range from tens to hundreds of square meters. This severely limits the form and composition of the substrates that can be used. In partial compensation for these difficulties, thermodynamic techniques have a number of advantages. They are versatile, the same apparatus being suitable for many combinations of gas and substrate over wide ranges of temperature. The methods are in some regimes extremely sensitive to small changes in the state of the films or substrate, rivaling most of the newer methods in sensitivity. Recent work on vapor pressures of monolayer and multilayer films have shown significant improvements in the state of the art (46–48). Heats of adsorption are currently measured by conventional methods (49, 50) and by several variants which study nonequilibrium effects; examples of these are flash desorption (51) and photodesorption (52). Modern improvements in instrumentation of calorimetry have made specific heat studies into a high resolution probe of adsorption (53), particularly when combined with vapor pressure measurements (54).

3.2 ADSORPTION PROPERTIES OF CLEAN UNIFORM SURFACES

The complete characterization of an adsorption substrate requires an extensive list of parameters whose relative importance depends upon the particular processes and system under study. For most purposes a small number of parameters are quite adequate for a basic description. First in importance to the majority of studies is the single adatom binding energy or the closely related heat of adsorption at low coverage. Such measurements have been made using various techniques on several combinations of gases with uniform substrates. The experiments show that the binding is highly specific to the type of gas atom, the substrate composition, and the crystal plane exposed for adsorption. Table 3.1 lists the binding energies of Ar, Kr, and Xe on the individual planes of a single crystal of tungsten, as measured by Engel and Gomer (20) by field emission techniques. The variations, both with respect to the gas atom and also crystal plane, are com-

Table 3.1 Binding Energies of Ar, Kr, and Xe on W Planes[a]

Plane	Binding energy (cal/mole)					
	Ar		Kr		Xe	
(110)[b]	2160	3250	7350	12,000	9200	11,600
(211)	0.87	0.76	0.75	0.65	0.71	0.5
(210)	0.99	0.96	0.77	0.67	0.70	0.5
(100)	0.87	0.75	0.74	0.67	0.64	0.56
(111)	0.89	0.77	0.69	0.61	0.64	0.43
N_1	7.9×10^{14}	2×10^{14}	7.2×10^{14}	2×10^{14}	6×10^{14}	2×10^{14}

[a] From Engel and Gomer (20).

[b] The entries under (110) refer to the heat of absorption on that plane for the assumed monolayer values shown in the line labeled N_1, given as fractions of the (110) plane binding.

parable to the theoretical estimates discussed in Chapter 2. Ehrlich and collaborators have also measured binding energies, by field desorption and by time and temperature studies with field emission observations of noble gas atoms on single crystal surfaces of tungsten (55), molybdenum (56), and rhodium (57); all show a high degree of specificity. Further examination of the data raises a few questions. The binding energies (see Table 3.1) vary by more than a factor two between the largest and smallest values measured for Xe; for Kr the total variation is smaller, and smaller still for Ar. Therefore, the trend observed seems to be toward lower specificity for more weakly bound gases, which is consistent with early opinions that since for weakly bound physisorbed gases the interaction is summed over relatively large numbers of substrate atoms, the binding should be relatively insensitive to surface structure. However, Engel and Gomer found that the order of preference is inconsistent with pair summations of van der Waals forces over the immediate neighbors of an adatom. If dispersion forces were dominant, adatoms would prefer "rough" planes on which adsorption sites are partially embedded in the surface; the order of preference would be (116) > (130) = (120) > (100) > (110)

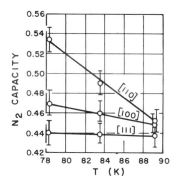

FIG. 3.3   N$_2$ monolayer capacities of individual crystal faces of copper measured by Rhodin (58).

(55), while Engel and Gomer found (110) significantly greater than (100) for the three adsorbates. Of course we do not now find it surprising that pairwise summation should fail with metallic substrates, but it is not clear whether the observed order of preference is consistent with more sophisticated theories. As another surprise, the binding energies do not increase systematically from Ar to Kr to Xe, but instead show that Kr binding is highest on all planes. This suggests that the binding contains some elements of chemisorption character, since for pure dispersion forces the binding is proportional to the polarizability, which is greatest for Xe. It must also be kept in mind that the field desorption technique is not completely understood, and it is possible that the large fields employed affect the measurements. However, the strong dependence of binding on crystal face is not in doubt and has been measured by other, less perturbing methods. The first such demonstration was made by Rhodin (58), who deduced the heats of adsorption of N$_2$ on various crystal faces of Cu from vapor pressure measurements (see Fig. 3.3). Thus, although quantitative questions remain, there is no doubt of the importance of substrate structure to the binding energy of van der Waals molecules. It is probable that fine details, such as the electronic structure of the metal and of the adatom, are also relevant.

Field emission and field ionization techniques have yielded several

Table 3.2 Dipole Moments of Rare Gas Atoms at Low Coverage on W Planes[a]

	Ar		Kr		Xe	
Plane	μ (debyes)[b]	q/e[c]	μ (debyes)	q/e	μ (debyes)	q/e
(110)	1.15	0.09	7.0	0.54	5.0	0.36
(120)	0.60	0.12	1.55	0.12	1.8	0.13
(100)	1.23	0.10	0.36	0.10	2.9	0.21
(101)	0.63	0.05	1.8	0.14	1.2	0.18
(211)	1.15	0.09	1.58	0.12	2.4	0.09

[a] From Engel and Gomer (20).
[b] 1 debye = 10^{-18} esu $-$Å.
[c] Equivalent charge transfer in electron units, for a displacement of 0.7 Å.

additional single adatom characteristics on single crystals and all have some strong specificity with respect to crystal facet and gas atom. Table 3.2 lists the effective dipole moments of rare gas atoms on W planes, deduced from changes in the work function for electron emission. As discussed in Chapter 2, there is no satisfactory theory for induced moments, although recent suggestions have opened new avenues of investigation. Atomic mobility is directly visualized by FEM, and a strong dependence on direction is seen. Anisotropy of mobility is to be expected on the basis of the regular atomic structures of the crystal surfaces, and it is gratifying to find these expectations confirmed. There are inconsistencies, however, with the predictions based upon simple models; for example, it is found that the self-diffusion of single W atoms on W surfaces is not systematically faster on smoother faces (59). The anomalies apparently involve the correlated motions of pairs of atoms (60).

The foregoing are samples of the highly detailed information that can be obtained using field emission and other microscopic techniques. There are many unsettled questions as to interpretation, however, and these will probably require a number of years before they are resolved. Nevertheless, it seems clear that the microscopic properties of certain real adsorbing surfaces can be observed in nearly ideal conditions, permitting quantitative comparisons with theoretical models.

3.3 HETEROGENEITY OF TYPICAL ADSORBENTS

A most important part of the description of any substrate involves its uniformity: how does it depart from the adsorption characteristics of an ideal surface, i.e., a flat, atomically clean facet of large single crystal? In nearly all circumstances the heterogeneity of the substrate plays a considerable role in the behavior of its adsorbed films. The evidence for the heterogeneity of typical adsorbents is reviewed in this section and the next. We return to the subject of heterogeneity in Chapter 9, where the emphasis is on the theory of films on such surfaces, utilizing a number of models worked out in intervening chapters. Therefore, although the emphasis here is rather on the experimental observations of heterogeneity, there is a certain amount of duplication with topics in Chapter 9.

One of the most direct tests of substrate uniformity is provided by a vapor pressure isotherm in the region of low coverages, where all gases should tend toward "Henry's law," i.e., vapor pressures proportional to coverage. Henry's law is usually derived as the limiting behavior of particular models, e.g., classical mobile and localized films, and it is sometimes quoted only in those contexts, but it is the asymptotic law for all films on uniform surfaces. It is only necessary that the coverage be sufficiently low to allow the neglect of adatom–adatom interactions (and such a regime exists at finite T even if the film displays phase condensation at higher coverage), and in this regime the only remaining interactions are those between single adatoms and the substrate. In these circumstances the most general formulation of the problem is that of the band model of noninteracting atoms, which yields Henry's law at low densities (see Chapter 5) for bosons and fermions regardless of bandwidths and energy gaps. The linear dependence is also unaffected by substrate distortion, for even if there are appreciable substrate-mediated interactions, they will always have some finite range and hence can be neglected when the interparticle distances are substantially greater than this range. Practical applications of Henry's law as a test of homogeneity must therefore be made with some consideration for the interactions in the particular system under examination. As a rough guide for typical gases and substrates, the linear region should be seen at coverages below about 0.1 monolayer, i.e., where interparticle distances are greater than three adatom diameters.

When this simple test is applied to experimental systems, one finds that heterogeneity is the rule; except for a relatively small number of studies on a smaller number of different substrates, low coverage isotherms display significant departures from Henry's law. These departures are always of the same general form: as coverage is reduced, the pressure falls more rapidly than the coverage indicating that the remaining adatoms are bound more strongly. For most of the systems reported in the vast literature on physical adsorption there is no linear region, but for others, Henry's law is obeyed over a range of coverage above some small value; in these cases there appears to be a limited fraction of the surface contributing to the variation, the remainder of the substrate being uniform to within the sensitivity of the measurement. If the surface is highly disordered then the heterogeneity will, of course, dominate over adatom interactions even at quite substantial coverages; in such cases the isotherm will not be linear over any region, but will instead be concave toward the pressure axis over a wide range of coverage.

Calorimetric measurements of heats of adsorption have given direct evidence for the rule of heterogeneity. The heat of adsorption is approximately equal to the single adatom binding energy, differing only by an added term $\simeq kT$ for weakly interacting films (see Chapters 4 and 5). On a very uniform surface the heat of adsorption is nearly independent of coverage until the layer approaches maximum density, but on a heterogeneous surface the heat of adsorption decreases as more atoms are adsorbed. This is simply due to the preferential adsorption of gas on the more "energetic" parts of the surface first, those having the greatest binding energy, and then on progressively less attractive regions. A typical curve for a strongly heterogeneous ad-

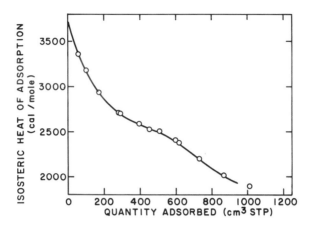

FIG. 3.4   Heat of adsorption of Ar on TiO_2 powder, by Drain and Morrison (61). The steep decrease with increasing coverage is characteristic of strongly heterogeneous surfaces.

sorbent is shown in Fig. 3.4, which reprints the measurements of Drain and Morrison of Ar adsorption on TiO_2 (61). In contrast, Fig. 3.5 displays the heat of adsorption of Xe on a relatively uniform surface, the (100) plane of a Pd single crystal (9).

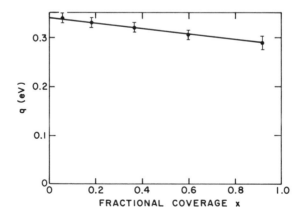

FIG. 3.5   Heat of adsorption of Xe on the (100) face of a palladium crystal measured by Palmberg (9). The nearly constant value indicates a very uniform surface.

One of the principal sources of heterogeneity in substrates such as TiO$_2$ powder is the exposure of several types of crystal plane for adsorption. Polycrystalline surfaces are heterogeneous to adsorption but the converse is not necessarily true; clean single crystal surfaces may fail to show homogeneous adsorption for various reasons. Two or more distinct binding energies have been observed in some cases of chemisorption (62), where it is ascribed to different ionization states of the adsorbate. Multistate adsorption may also occur in physisorption of simple atoms if the surface contains two or more distinct classes of adsorption sites (63). Even when the crystal structure of a surface plane contains sites of only one class, it will in practice contain a number of growth steps, vacancies, and dislocations. These imperfections, although not generally distinguishable by microscopic probes such as LEED, will introduce a degree of heterogeneity in adsorption (64). Adsorption may even be affected by deeper imperfections which are not evidenced on the surface; extensive subsurface disorder has been observed by high energy reflection electron microscopy (65).

Chemical contamination still remains as one of the most common sources of heterogeneity. The presence of strongly bound impurities on the surface can cause large and, in some cases, unpredictable effects. For example, Engel and Gomer (66) found that chemisorbed oxygen caused marked reductions in the binding of Kr and Xe to W, while Halpern and Gomer (67) observed that oxygen increased the binding of He on W. A surface is likely to be contaminated before an experiment begins; it may even occur as a byproduct of the study. This is particularly likely when the substrate is colder than other portions of the apparatus which are in the same vapor space, even under ultrahigh vacuum conditions.

3.4 PREPARATION OF UNIFORM SURFACES

Techniques of surface preparation include heating in vacuum, cleaving, cutting, polishing and etching, vapor deposition, ion bombardment, and field desorption. Several of these are suitable in only certain types of investigation, e.g., field desorption is restricted to FEM and

FIM studies, and ion bombardment for relatively flat surfaces completely open to the ion discharge.

Other preparation techniques are less specific and more widely applied; heating in vacuum is most common. It is often the only technique, and while it may in certain cases be all that is necessary to prepare a manifestly uniform substrate, it is often quite inadequate. While moderate heating in vacuum can strip off weakly bound impurities, the temperatures at which adsorption systems are usually baked are too low to remove typical chemisorbed species. Some impurities are so strongly bound that they cannot be removed at temperatures below the melting point of the solid, as is the case for C on Cu (68). Heating may cause more harm than good, for it causes some bulk impurities to diffuse to the surface; extremely low impurity concentrations in the bulk are enough to supply considerable areal densities at the surface. Heating can also increase the number of imperfections and cause thermal etching of the surface.

Deposition from the vapor is often used in LEED work, where the resulting surface can be examined directly. Considerable success has been obtained with this technique, but there are also some limitations. LEED is relatively insensitive to growth steps and other surface imperfections: it provides a good view of average long-range structure but not of localized disturbances compatible with the overall order. LEED is also insensitive to strained epitaxial and amorphous growths of underlying layers, whereas they can cause perceptible changes in adsorption. A particular kind of vapor deposition has been found to improve the uniformity of some highly heterogeneous substrates. Steele and Aston (69) have shown that preadsorbed Ar can cause substantial improvements in the homogeneity of a TiO_2 adsorbent by preferentially coating the more strongly adsorbing sites. The method is not a panacea, however, for the heterogeneity of the base can be propagated through long-range interactions and also by the disorder it imposes on the preplated layers. A further danger is the possibility that precoating will actually increase the heterogeneity by forming patches and thereby introducing an entirely different type of surface. This has actually been observed by Singleton and Halsey (70).

The preparation of uniform surfaces from single crystals by cutting and polishing or by cleaving appear to be only partially successful (68).

Cutting and polishing may cause damage extending through hundreds of layers below the surface: such damage may be removed by etching, but etching can produce pits and leave chemisorbed impurities. Cleaving will generally leave a stepped surface and sometimes produces deeper damage.

The prototypical exceptions to the rule of heterogeneity are the graphitized carbon blacks. Attention was drawn to these materials by the work of Polley *et al.* (71, 72), who discovered that carbon black subjected to high temperature heat treatment displayed stepwise vapor pressure isotherms in argon adsorption. The stepwise isotherms were taken as evidence of distinct layer formation, approaching the ideal forms predicted theoretically (see Chapter 9). The experimental isotherms on a series of carbon blacks showed steps which became more distinct on those samples which had been heated to higher temperatures, and this evidence of increasing homogeneity was correlated with x-ray indication of the progressive development of graphite structure (73). The x-ray evidence alone could not account for the adsorption homogeneity, however, since binding energies on basal planes, edge planes, and imperfections are different (74, 75). An additional essential feature was deduced from electron microphotographs which reveal that the high temperature causes the particles, originally nearly spherical in shape, to grow in size and to develop flat external surfaces with symmetries corresponding to basal planes. It therefore appears that the graphitized blacks have only basal planes exposed for adsorption. As corroboration, it has been recently found that $FeCl_3$, which is readily intercalated into natural and pyrolytic graphite crystals by diffusion through edge planes, will not readily enter highly graphitized carbon black (46). The most striking confirmation of these deductions has been obtained in very high resolution electron microscopy of carbon blacks in various stages of progressive graphitization, in which the individual crystal planes become resolved, and these show that the highly ordered particles have only basal plane surfaces exposed for adsorption (76, 77) (Fig. 3.6). The evident adsorption homogeneity of the graphitized blacks is also due to the refractory nature of graphite, which permits heating to high enough temperatures to remove most contaminants.

Graphitized carbon black is virtually unique among adsorbents, but

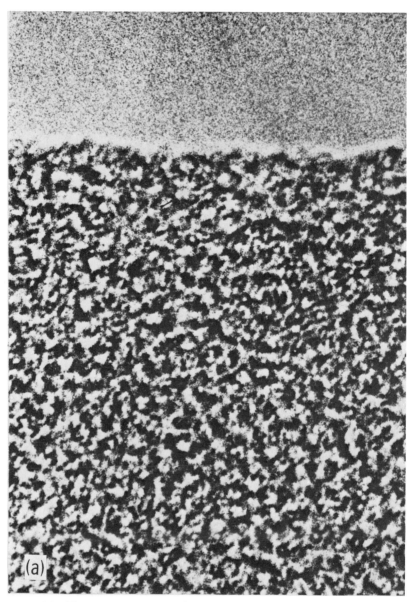

FIG. 3.6 Electron micrographs of amorphous and graphitized carbon by Ban and Hess (76, 77), using high resolution diffracted beam electron microscopy. The material in (a) is a carbon black obtained by pyrolysis of an organic compound at relatively low temperature. It shows no regular crystal structure. On heating to temperatures above 1000°C. graphitization begins to occur; prolonged heating at high temperature produces complete graphitization of the exposed surfaces, as seen in (b). Here the resolution is sufficient to show the lattice image from the (100) reflection with d spacing of 3.35 Å. (Courtesy of L. L. Ban.)

(b)

51

52

a related material—exfoliated graphite—has been found to rival its homogeneity.† Exfoliation involves the formation of intercalation complexes within layer-structured crystals and subsequent sudden heating which explodes the layers apart. Intercalation complexes can be formed with many crystals and inorganic and organic chemicals (78); included among the substances which penetrate graphite are alkali metals, halogens, and trivalent chlorides of Fe, Co, Cu, and Al (79). Their maximum concentrations correspond to one guest molecule for alternate basal plane hexagons of the graphite host. Exfoliation by heating causes very large increases in the basal plane areas exposed for adsorption but no corresponding increase in the areas of edge planes. The heating removes some of the guest molecules, but substantial traces of some components (e.g., transition metal ions) may remain. Despite the heterogeneity which must be due to edge plane areas and chemical residues, the adsorption characteristics of exfoliated graphite are somewhat superior to highly graphitized carbon black. Thomy and Duval (46) have made detailed vapor pressure studies of Kr, Xe, and

† A particularly useful form of exfoliated graphite is available commercially, marketed under the trade name "Grafoil" by Union Carbide Corporation, Carbon Products Division, 270 Park Avenue, New York. It is composed of graphite particles exfoliated in a strongly oxidizing medium such as sulfuric acid, then rinsed and rapidly heated. After heat treatment, the expanded particles are pressed together and rolled into binderless flexible sheets with density about one-half that of crystalline graphite. The specific adsorption area is about 20 m^2/gm. The basal plane surfaces are strongly oriented parallel to the plane of the sheet; neutron diffraction of a representative sample indicates a distribution of orientations about the surface normal, with full width at half maximum of about 30°(45). Chemical purity is high, with typical concentrations of metal ions (Al, Cu, Fe) of the order of 10 ppm. The graphitized blacks were originally produced by the Cabot Corp., Boston, Massachusetts, and material similar to the grades studied earlier are still obtainable from this company. The parent material is designated as Spheron, while the more homogeneous highly graphitized blacks have been described under various names which usually include reference to the temperature of heat treatment, e.g., Spheron (2700°) and Graphon (2700°). The latter is sometimes referred to in the scientific literature simply as Graphon, or by the batch number, e.g., Graphon P-33 (2700°) = P-33.

FIG. 3.7   Scanning electron micrograph of a cleaved surface of Grafoil, exhibiting the preferential basal plane alignment of the crystallites. The linear dimensions of the hexagonal segments range from about 1000 to 3000 Å. (Photograph taken at the Metallurgy Department, AEK RISØ, Denmark. Courtesy of Jørgen Kjems.)

methane on several samples of natural, pyrolytic, and SiC-derived graphites exfoliated after the intercalation of $FeCl_3$, and they compared these substrates with graphitized carbon black (3000°) and with natural graphite. For each of the gases the isotherms showed appreciably sharper steps on exfoliated graphite than on the other substrates. Furthermore, isotherms on exfoliated graphite display novel "substep" details which had not been seen previously in studies with the graphitized blacks (the substeps, attributed to liquid–solid transitions in a single layer, are discussed in subsequent chapters). Duval and Thomy attribute the apparent superiority of exfoliated graphite to the larger areas of the individual adsorbing surfaces, postulating that the small grain size of graphitized black produces some degree of capillary condensation. The particular form of exfoliated graphite known as Grafoil (see footnote above) has been used as the substrate in several studies of adsorption, and in each case, the experimental results indicate exceptional uniformity. However, there are definite signs of appreciable variation in binding energy over a small fraction of the surface. According to the careful vapor pressure measurements of Elgin and Goodstein (54) the single particle binding energy of ^{1}He has the empirical distribution function

$$f(E) = 0.045[-(E + 143)]^{-4/3} \qquad (3.4.1)$$

where $143.0025 < -E < 268$, with E in units of k_B. According to Eq. (3.4.1), 95% of the surface has binding energy between 143 k_B and 149 k_B, and 90% has binding within 1.% of 143 k_B.

A limited number of other adsorbents display features of distinct layer formation, approaching but not quite equaling the evident adsorption uniformity of exfoliated graphite. Such materials yielding "semistepwise" isotherms include MoS_2 (80, 81); divalent metal–halide compounds (47); BN (83), NaCl, KCl, RbCl (48, 83), and NaBr (84). The divalent metal–halide compounds have layer structures similar to graphite and hence their uniformity is probably due to the dominance of basal plane surfaces exposed for adsorption. Since BN and MoS_2 show some willingness to form intercalation complexes (79), it is possible that these substances might be exfoliated by procedures similar to those applied to graphite, with comparable improvements in adsorption homogeneity.

This outline is primarily intended as a reminder of the difficulties of surface preparation and not as a guide to techniques. No single technique can be applied to all types of surface, but should be used with attention to the specific adsorbent, adsorbate, and experimental probe. In the final analysis, the data from any series of experiments must still be treated as if heterogeneity were present; surface uniformity can only be indicated by the testing of results against other experiments on similar systems and by comparisons with theoretical models of homogeneous and heterogeneous adsorption.

REFERENCES

1. J. P. Hobson, *Proc. Int. Vacuum Congr., 6th, Kyoto 1974.*

2. T. W. Haas *et al., Progr. Surf. Sci.* **1,** p. 155. (1971)

3. J. J. Lander, *Recent Progr. Solid State Chem.* **2,** 26 (1965).

4. M. B. Webb and M. G. Lagally, *Solid State Phys.* **28,** 302 (1973).

5. C. B. Duke and G. E. Laramore, *Phys. Rev.* **B3,** 3183 (1971).

6. R. M. Stern, J. J. Perry, and D. S. Boudreaux, *Rev. Mod. Phys.* **41,** 275 (1969).

7. H. H. Farell, M. Strongin, and J. M. Dickey, *Phys. Rev.* **B6,** 4703 (1972).

8. J. J. Lander and J. Morrison, *Surface Sci.* **6,** 1 (1967).

9. P. W. Palmberg, *Surface Sci.* **25,** 598 (1971).

10. J. J. Lander, *Phys. Rev.* **91,** 1382 (1953).

11. L. A. Harris, *J. Appl. Phys.* **39,** 1419 (1968).

12. J. C. Tracy, Jr., *in* "Electron Emission Spectroscopy" (W. Dekeyser *et al.,* eds.). Reidel, Dordrecht, The Netherlands, 1973.

13. J. Suzanne, J. P. Coulomb, and M. Bienfait, *Surface Sci.* **40,** 414 (1973); **44,** 141 (1974.).

14. K. Siegbahn *et al.,* "Atomic Molecular and Solid State Structure Studied by Electron Spectroscopy." Almquist and Wiksalls, Uppsala, 1967.

15. N. J. Taylor, "Techniques of Metals Research," (R. F. Bunshah, ed.), Vol. 7. Wiley (Interscience), New York.

16. H. D. Hagstrum, *Phys. Rev.* **150,** 495 (1966).

17. H. D. Hagstrum, *Science* **178,** 275 (1972).

18. R. Gomer, "Field Emission and Field Ionization." Harvard Univ. Press, New Haven, Connecticut, 1969.

19. E. W. Müller and T. T. Tsong, "Field Ion Microscopy." American Elsevier, New York, 1969.

20. T. Engel and R. Gomer, *J. Chem. Phys.* **52,** 5572 (1970).

21. J. W. Gadzuk and E. W. Plummer, *Rev. Mod. Phys.* **45**, 487 (1973).

22. A. J. Melmed and E. W. Müller, *J. Chem. Phys.* **29**, 1037 (1958).

23. G. Ehrlich and C. F. Kirk, *J. Chem. Phys.* **48**, 1465 (1968).

24. T. T. Tsong, *Phys. Rev. Lett.* **31**, 1207 (1973).

25. E. W. Müller, J. Panitz, and S. B. McLane, *Rev. Sci. Instrum.* **39**, 83 (1968).

26. R. Frisch and O. Stern, *Z. Phys.* **84**, 430 (1933).

27. J. E. Lennard-Jones and A. F. Devonshire, *Nature (London)* **137**, 1039 (1936).

28. R. E. Stickney, *Advan. At. Mol. Phys.* **3**, 143 (1967).

29. J. N. Smith, Jr. and H. Saltsburg, *in* "Fundamentals of Gas-Surface Interactions" (H. Saltsburg, J. N. Smith, Jr., and M. Rogers, eds.), Academic Press, New York, 1967.

30. B. F. Mason and B. R. Williams, *J. Chem. Phys.* **56**, 1895 (1972).

31. W. H. Weinberg and R. P. Merrill, *Phys. Rev. Lett.* **25**, 1198 (1970).

32. D. E. Houston and D. R. Frankl, *Phys. Rev. Lett.* **31**, 298 (1973).

33. P. Drude, *Ann. Phys.* **272**, 532, 865 (1889); **275**, 481 (1890).

34. N. M. Bashara, A. B. Buckman, and A. C. Hall (eds.), *Proc. Symp. Recent Develop. Ellipsomet., Surface Sci.* **16** (1969).

35. A. J. Melmed, "Molecular Processes on Solid Surfaces" (E. Drauglis, R. D. Gretz, and R. I. Jaffee, eds.), McGraw-Hill, New York, 1969.

36. J. G. Aston, *in* "The Solid-Gas Interface" (E. A. Flood, ed.), Vol. 1. Dekker, New York, 1966.

37. R. J. Rollefson, *Phys. Rev. Lett.* **29**, 410 (1972).

38. R. J. Rollefson, *in Proc. Symp. Submonolayer He Films* (J. G. Daunt and E. Lerner, eds.). Plenum Press, New York, 1973.

39. D. J. Creswell, D. F. Brewer, and A. L. Thomson, *Phys. Rev. Lett.* **29**, 1144 (1972).

40. J. W. Riehl and C. J. Fisher, *J. Chem. Phys.* **59**, 4336 (1973).

41. V. I. Goldanskii and R. H. Herber, "Chemical Applications of Mössbauer Spectroscopy." Academic Press, New York, 1968.

42. M. J. D. Low, *in* "The Solid-Gas Interface" (E. A. Flood, ed.), Vol. 2. Dekker, New York, 1966.

43. S. Bukshpan and S. L. Ruby, *Bull. Amer. Phys. Soc. (Ser. 2)* **16**, 850 (1971); S. Bukshpan, private communication.

44. J. K. Kjems, L. Passell, H. Taub, and J. G. Dash, *Phys. Rev. Lett.* **32**, 724 (1974).

45. J. K. Kjems, L. Passell, H. Taub, J. G. Dash, and A. D. Novaco, *Phys. Rev.*

46. A. Thomy and X. Duval, *J. Chim. Phys. Physicochim. Biol.* **66**, 1966 (1969); **67**, 286, 1101 (1970).

47. Y. Larher, *J. Phys. Chem.* **72**, 1847 (1968); *J. Colloid Interface Sci.* **37**, 836 (1971).

48. T. Takaishi and M. Mohri, *J. Chem. Soc., Faraday Trans. I* **68**, 1921 (1972).

49. E. L. Pace, "The Solid–Gas Interface" (E. A. Flood, ed.), Vol. 1, Chapter 4. Dekker, New York, 1966.

50. J. M. Holmes, Ref. 49, Chapter 5.

51. M. Abon, B. Tardy, and S. J. Teichner, "Adsorption-Desorption Phenomena" (F. Ricca, ed.), p. 245. Academic Press, 1972.

52. P. Kronauer and D. Menzel, Ref. 51, p. 313.

53. M. Bretz, J. G. Dash, D. C. Hickernell, E. O. McLean, and O. E. Vilches, *Phys. Rev.* **A8**, 1589 (1973); **A9**, 2814 (1974).

54. R. L. Elgin and D. L. Goodstein, *Phys. Rev.* **A9**, 2657 (1974).

55. G. Ehrlich and F. G. Hudda, *J. Chem. Phys.* **30**, 493 (1959).

56. G. Ehrlich, *Brit. J. Appl. Phys.* **15**, 349 (1964).

57. G. Ehrlich, H. Heyne, and C. F. Kirk, *in* "The Structure and Chemistry of Solid Surfaces" (G. A. Somorjai, ed.). Wiley, New York, 1969.

58. T. N. Rhodin, Jr., *J. Amer. Chem. Soc.* **72**, 5691 (1950).

59. G. Ehrlich and F. G. Hudda, *J. Chem. Phys.* **44**, 1039 (1966).

60. W. R. Graham and G. Ehrlich, *Phys. Rev. Lett.* **31**, 1407 (1973).

61. L. E. Drain and J. A. Morrison, *Trans. Faraday Soc.* **48**, 316 (1952).

62. See, for example, J. Anderson and P. J. Estrup, *J. Chem. Phys.* **46**, 563 (1967).

63. See, for example, S. Ross and H. Clark, *J. Amer. Chem. Soc.* **76**, 4291, 4297 (1954).

64. W. J. Dunning, "The Solid-Gas Interface" (E. A. Flood, ed.), Vol. 1. Dekker, New York, 1967.

65. B. M. Siegel and J. F. Menadue, *Surface Sci.* **8**, 206 (1967); J. F. Menadue and B. M. Siegel, *in* "The Structure and Chemistry of Solid Surfaces" (G. A. Somorjai, ed.), Wiley, New York, 1969.

66. T. Engel and R. Gomer, *J. Chem. Phys.* **52**, 5572 (1970).

67. B. Halpern and R. Gomer, *J. Chem. Phys.* **51**, 3043 (1969).

68. H. E. Farnsworth, "The Solid-Gas Interface" (E. A. Flood, ed.), Dekker, New York, 1966.

69. W. A. Steele and J. G. Aston, *J. Amer. Chem. Soc.* **79**, 2393 (1957).

70. J. H. Singleton and G. D. Halsey, Jr., *J. Phys. Chem.* **58**, 330, 1011 (1954).

71. M. H. Polley, W. D. Schaeffer, and W. R. Smith, *J. Phys. Chem.* **57**, 469 (1953).

72. W. D. Schaeffer, W. R. Smith, and M. H. Polley, *Ind. Eng. Chem.* **45**, 172 (1953).

73. R. E. Franklin, *Acta Cryst.* **4**, 253 (1951).

74. R. M. Barrer, *Proc. Roy. Soc.* **A161**, 476 (1937).

75. E. F. Meyer and V. R. Deitz, *J. Phys. Chem.* **71**, 1521 (1967).

76. W. M. Hess and L. L. Ban, *Proc. Int. Congr. Electron Microsc., 6th, Kyoto* **1**, 569 (1966).

77. L. L. Ban and W. M. Hess, *Proc. Bicenten. Conf. Carbon, 9th, Boston* 1969.

78. R. M. Barrer, *in* "Non-Stoichiometric Compounds" (L. Mandelcorn, ed.), Chapter 6. Academic Press, New York, 1964.

79. G. R. Hennig, *Progr. Inorg. Chem.* **1**, 125.

80. L. Bonnetain, X. Duval, and M. Letort, *C. R. Acad. Sci. Paris* **234**, 1363 (1952).

81. E. V. Ballou, *J. Amer. Chem. Soc.* **76**, 1199 (1954).

82. R. A. Pierotti, *J. Phys. Chem.* **66**, 1810 (1962).

83. S. Ross and H. Clark, *J. Amer. Chem. Soc.* **76**, 4291 (1954).

84. B. B. Fisher and W. G. McMillan, *J. Amer. Chem. Soc.* **79**, 2969 (1957).

4. The Statistical Thermodynamics of Physisorption

The formal development of adsorption thermodynamics began with Gibbs's great work, "On the Equilibrium of Heterogeneous Substances" (1). Subsequent contributions to the theory were made by many people, and a short list of some papers that continue to be cited with regularity is given at the end of the chapter (2–9). This abbreviated list only hints at what is a long and very extensive literature,† but nevertheless, the thermodynamics of adsorption does not yet appear to be a completely closed subject. The lack of a precise boundary between film and vapor, the strong gradients in film properties near the solid surface, and the difficulties of putting theory to experimental tests, are persistent problems contributing to continuing discussion and controversy. These questions are made more troublesome by the lack of a formalism and notation consistent within the field. Separate styles of notation is typical of thermodynamics, but such individuality seems exaggerated in the case of adsorption thermodynamics. Here one even finds disagreements as to the proper formulation of basic thermodynamic functions, as, for example, the Helmholtz and Gibbs free energies. In this situation one must return to fundamentals, inspecting the basic connections between thermodynamic and funda-

† For a much more complete review of the literature, see Young and Crowell (10).

mental statistical quantities to develop a general thermodynamic theory of adsorption beginning with statistical mechanics. This approach offers several advantages: it provides clear connections with other branches of statistical thermodynamics, it yields unequivocal definitions of proper free energies, it simplifies the derivation of many "standard" formulas, and it provides a natural formalism for microscopic models of monolayers and multilayers. This is not an entirely novel approach to surface thermodynamics; Guggenheim (4), for example, strongly advocated a statistical view. Perhaps its most familiar exposition is the chapter on surface layers in the text by Fowler and Guggenheim (11). But whereas Fowler and Guggenheim focus their interest on a few special models of monomolecular films, we derive general thermodynamic relationships which serve as a general framework suitable for any model. For the most part, we use the general style and notation of Landau and Lifshitz (12), with the usage of the conventional nomenclature of adsorption wherever it is well established and not in conflict with the wider field of thermodynamics.

4.1 THE COMPLETE ADSORPTION SYSTEM AND ITS DIVISION INTO SEPARATE COMPONENTS

Consider a vessel containing definite quantities of adsorbent (which will be equivalently termed "substrate") and adsorbate. It is not necessary at this stage to define these terms, but we anticipate later specification and take their meanings qualitatively. "Substrate" refers to a relatively stable solid material having an appreciable surface area, and "adsorbate" to a mixture of vapor and its film. Although in most cases of physical interest the thermodynamic properties of adsorbent and adsorbate appear clearly separable, their distinction cannot in principle be exact, and therefore it will be useful to develop the theory, at the outset at least, with greater generality. A division of the adsorbate between film and vapor is even more diffuse. We shall therefore begin by considering substrate, film, and vapor as distinctions in name only, the thermal properties of the system being due to a mixture of all three. Their later subdivision into components having separable properties will come about only after the adoption of specific physical assumptions.

The mixed system will in general have a minimum energy E_0 and a set of higher levels E_i with level density $\Gamma(E)$. In thermal equilibrium with a reservoir at temperature T, the probability that the system is in a state with energy E_i is

$$\omega_i = \exp(-\beta E_i)/\sum_i e^{-\beta E_i}, \qquad \sum_i \omega_i = 1, \qquad (4.1.1)$$

where $\beta = (kT)^{-1}$. The canonical partition function Z for a fixed number of particles in the total system is the sum over states

$$Z = \sum_i \exp(-\beta E_i) = \int_{E_0}^{\infty} e^{-\beta E}\Gamma(E)\, dE. \qquad (4.1.2)$$

Now Z can be written formally as the product of separate terms for film, vapor, and substrate:

$$Z = Z_f Z_v Z_s. \qquad (4.1.3)$$

This factoring does not imply any special properties or distinguishability of the parts. It is always permitted by the convolution property of the density of states: for any composite system $A + B$,

$$\Gamma_{A+B}(E) = \int \Gamma_A(E')\Gamma_B(E - E')\, dE'. \qquad (4.1.4)$$

The statistical properties of film, vapor, and substrate can be formally related through the factoring of the total Z and the familiar general results of statistical mechanics. The derivations are outlined in the following.

Energies

The ensemble average energy E of the total mixed system is the weighted sum over states:

$$E = \sum_i \omega_i E_i. \qquad (4.1.5)$$

Substituting Eq. (4.1.1) for ω_i, we obtain

$$E = \frac{\Sigma_i \exp\ (-\beta E_i)E_i}{\Sigma_i \exp\ (-\beta E_i)} = -\frac{\partial(\ln Z)}{\partial\beta}. \qquad (4.1.6)$$

Writing Z in terms of the separate factors, Eq. (4.1.3), we see that E is the sum of component energies,

$$E = E_f + E_v + E_s, \qquad (4.1.7)$$

where

$$E_f = -\frac{\partial(\ln Z_f)}{\partial\beta}, \qquad E_v = -\frac{\partial(\ln Z_v)}{\partial\beta}, \qquad E_s = -\frac{\partial(\ln Z_s)}{\partial\beta}.$$

Entropies

The equilibrium entropy S of any system is related to the statistical distribution by

$$S = -\overline{k\ln\omega} = -k\sum_i \omega_i \ln \omega_i, \qquad (4.1.8)$$

which, with Eqs. (4.1.1), (4.1.2), and (4.1.6), yields

$$S = k\sum_i \omega_i(\beta E_i + \ln Z) = E/T + k\ln Z. \qquad (4.1.9)$$

Using the additivity of energy equation (4.1.7) and Eq. (4.1.3) for Z, S is the sum

$$S = S_f + S_v + S_s, \qquad (4.1.10)$$

where

$$S_f = E_f/T + k\ln Z_f, \qquad (4.1.11)$$

and similarly for vapor and substrate.

Helmholtz free energy

The fundamental connection between statistical and thermodynamic properties of the canonical ensemble is through the Helmholtz free energy F:

$$F = -kT\ln Z. \qquad (4.1.12)$$

With the factoring of Z we immediately obtain the free energies of the components

$$F = F_f + F_v + F_s,\qquad(4.1.13)$$

where

$$F_f = -kT \ln Z_f,\ \text{etc.}\qquad(4.1.14)$$

Combining relations of the type (4.2.4) with (4.2.1), we obtain connections between energy, entropy, and free energy of the total system and of each component:

$$F = E - TS,\qquad(4.1.15)$$

$$F_f = E_f - TS_f,\quad \text{etc.}\qquad(4.1.16)$$

Forces and displacements

The first law of thermodynamics for reversible processes involving generalized forces X_α and displacements x_α is

$$dE = dQ - \sum_\alpha X_\alpha\, dx_\alpha = T\, dS - \sum_\alpha X_\alpha\, dx_\alpha.\qquad(4.1.17)$$

With the expression for the total differential of energy obtained from Eq. (4.1.15), we have

$$dF = -S\, dT - \Sigma X_\alpha\, dx_\alpha.\qquad(4.1.18)$$

Equation (4.1.18) is perfectly general for any type of system, whether homogeneous or composed of different parts as is the adsorption system. It is at this point that we can effect a definite separation between the thermal properties of film and vapor. We recognize that the area of the film must in some sense play a role analogous to volume of the vapor. Both are extensive coordinates, i.e., are generalized displacements x. In the case of the volume V the conjugate generalized force is the pressure P; for the film the force conjugate to the area A is called the spreading pressure and conventionally symbolized as ϕ or π (see Fig. 4.1). With these definitions, the two pressures can be expressed as partial derivatives of the total free energy of the composite system: from Eq. (4.1.18), we obtain

$$P = -(\partial F/\partial V)_{T,A},\qquad \phi = -(\partial F/\partial A)_{T,V}.\qquad(4.1.19)$$

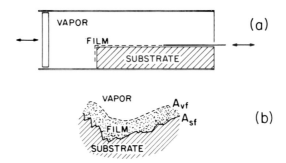

FIG. 4.1   (a) Components of an adsorption system, illustrating a volume compression of the vapor by an ordinary 3D piston, and an areal compression of the film by a surface barrier. To a first approximation, a thin film is assumed to be strictly 2D, but diagram (b) shows that the actual situation can be much more complicated. Gradients in multilayer films are discussed in Section 4.6.

Equation (4.1.19) implies at least a partial separation of the adsorbate into film and vapor fractions. We can certainly expect that the principal contribution to P will be due to the vapor; we also expect that the major term in ϕ is due to the film. Thus, if we express F in terms of its parts, (4.1.19) yields

$$P = -(\partial F_v/\partial V)_{T,A} - (\partial F_f/\partial V)_{T,A} - (\partial F_s/\partial V)_{T,A},$$

$$\phi = -(\partial F_v/\partial A)_{T,V} - (\partial F_f/\partial A)_{T,V} - (\partial F_s/\partial A)_{T,V}. \quad (4.1.20)$$

These expressions, while correct in principle, would make subsequent calculations quite tedious if they had to be carried along in their entirety. While it must indeed be true that the 3D pressure P, for example, reflects some of the elastic responses of the substrate and film, these contributions are in all cases of interest much smaller than that due to the vapor. Similarly, it is the film term that is dominant in ϕ. In both cases we would wish to eliminate the minor terms completely, at least for all but the most exacting calculations. These reductions can be accomplished in two ways.

The more elegant alternative is to *define* film and vapor phases so as to effect a perfect separation of terms. Thus, since the division of the partition function into film, vapor, and substrate factors has been completely arbitrary to this point, we may here define the factors in whatever manner required to yield "pure" partial derivatives. In this way, the "film" would correspond to that portion of the system containing all of the variations on A but no dependence on V, and the "vapor" would depend on V but not at all on A. This method of separating film and vapor is effectively that of the "dividing surface" of Gibbs. Much of the subsequent theoretical developments in adsorption thermodynamics is based on this concept. But in spite of its long tradition, this artifice brings with it a number of serious problems. It effects too complete a division between film and vapor, so that a number of inconvenient constraints are placed on the interpretation of measurements and the construction of realistic theoretical models. For example, although there must be certain modifications produced in the equation of state of a vapor in the vicinity of a surface, these changes would have to be attributed to the "film" even though the interaction region actually might extend quite far into the vapor phase.

A less elegant or exact division of the adsorbate into film and vapor portions is to accept some initial approximations concerning their physical natures. As will be seen, these approximations will in their limiting cases turn out to create precisely the same separation as the "dividing surface" of Gibbs. But the advantage of this less elegant approach is that, just as in real adsorption systems, the separation between film and vapor is *not* exact, and it holds open the opportunity for some future, more faithful imitation of the actual situations. The assumptions are as follows:

(a)   The portion of vapor volume within the range of surface forces is a very small fraction of the total V.

(b)   The volume occupied by film is a very small fraction of V.

(c)   Any distortions or deformations of the substrate caused by adsorption are reversible continuous functions of the quantity of film adsorbed.

Assumptions (a) and (b) in their extreme forms, as already noted, yield "pure" expressions for ϕ and P, whereas they are in principle always composed of three terms, as Eq. (4.1.20). In the following sections we usually adopt the extreme forms but will also inspect the conditions under which they are satisfactory approximations.

Assumption (c) is actually a matter of definition and need not be inexact, depending upon subsequent models. It is a less restrictive and more physically reasonable assumption than that typically made concerning the nature of adsorbents. In conventional treatments it is assumed that the substrate is "inert," i.e., unaffected by adsorption. This assumption is unphysical in principle; furthermore, there are several known mechanisms by which the properties of a surface are appreciably modified in the presence of an adatom (13). Fortunately, one need not require the "inert substrate" assumption in order to proceed with the theory. It is only necessary that the substrate distortions be caused on an adatom-by-adatom basis, i.e., to be a single-valued function of the quantity of gas adsorbed on the surface. For if this is the case, then the perturbation is associated with the adsorption state of the adatom and "belongs" to the adatom just as intimately as its binding energy, dipole moment, or vibration amplitude. In fact, a proper calculation of the binding energy and other properties of the adsorption state must take into account the substrate distortions; as long as they can be related to the quantity adsorbed they may be treated as attributes of the adatoms themselves.

4.2 EQUILIBRIUM RELATIONS

The establishment of equilibrium in an adsorption system involves the flow of energy between substrate and adsorbate and an exchange of particles between film and vapor. Both of these processes can be analyzed in a variety of ways, depending on what parameters of the system are imagined to be held constant. These different situations corresponding to the microcanonical, canonical, and grand canonical ensembles, all yield the same results as to equilibrium properties; they differ only in their fluctuations.

Let us first consider the microcanonical ensemble, i.e., an isolated adsorption system having fixed total energy, volume, adsorption area, and number of particles. Equilibrium corresponds to the state of maximum entropy of the total system with respect to redistributions of energy and particles. Concerning energy flow, we first consider the system as composed of two parts; substrate (s) and adsorbate (a):

$$S = S_s + S_a = \text{const}, \tag{4.2.1a}$$

$$E = E_s + E_a = \text{const}. \tag{4.2.1b}$$

The entropy is maximized with respect to energy changes, so that at equilibrium

$$(\partial S/\partial E_s)_{N,V,A} = 0. \tag{4.2.2}$$

Using relations (4.2.1), we obtain

$$(\partial S_s/\partial E_s)_{N,V,A} = (\partial S_a/\partial E_a)_{N,V,A}. \tag{4.2.3}$$

These partial derivatives as can be seen from (4.1.17), are just the reciprocals of the absolute temperatures of substrate and adsorbate; thus Eq. (4.2.3) implies the equality of temperatures

$$T_s = T_a, \tag{4.2.4}$$

and the result can obviously be extended to any two portions of the system: T is then uniform throughout.

Now we consider the exchange of particles between film and vapor: at equilibrium the entropy is a maximum, so that

$$(\partial S/\partial N_f)_{E,V,A} = 0. \tag{4.2.5}$$

Now we must consider all three parts of the total entropy

$$S = S_f + S_v + S_s,$$

so that the partial derivative in (4.2.5) becomes

$$(\partial S_f/\partial N_f)_{E,V,A} + (\partial S_s/\partial N_f)_{E,V,A} - (\partial S_v/\partial N_v)_{E,V,A}. \tag{4.2.6}$$

The second term would be awkward to carry through subsequent manipulations, but fortunately it vanishes according to assumption (c), since all substrate changes caused by adsorption are to be associated with the adatoms: such changes are actually embedded in the first term, $(\partial S_f/\partial N_f)$. Thus, we have the fundamental equation of vapor–film equilibrium,

$$(\partial S_f/\partial N_f)_{E,V,A} = (\partial S_v/\partial N_v)_{E,V,A}. \tag{4.2.7}$$

The quantities in Eq. (4.2.7) involve the chemical potentials of the film and vapor. To show this, we note that the total differentials dE and dF in Eqs. (4.1.17) and (4.1.18) must actually include terms proportional to the quantity of material: in addition to the products of conjugate forces and displacements $\phi \, dA$ and $P \, dV$, the sums $\Sigma \, X_\alpha d \, X_\alpha$ also include products involving dN. Now, if we introduce this explicit dependence in Eq. (4.1.17) and also write out the separate contributions to the energy, entropy, etc., we obtain

$$dE = dE_{\mathrm{f}} + dE_{\mathrm{v}} + dE_{\mathrm{s}}$$

$$= T(dS_{\mathrm{f}} + dS_{\mathrm{v}} + dS_{\mathrm{s}}) - \phi \, dA - P \, dV$$
$$+ \mu_{\mathrm{f}} \, dN_{\mathrm{f}} + \mu_{\mathrm{v}} \, dN_{\mathrm{v}} + \mu_{\mathrm{s}} \, dN_{\mathrm{s}} . \qquad (4.2.8)$$

The μ's are introduced here simply as the generalized coordinates conjugate to particle number. They are directly related to certain partial derivatives of the energies and entropies: differentiating with appropriate quantities being held constant, we obtain

$$\mu_{\mathrm{f}} = (\partial E_{\mathrm{f}} / \partial N_{\mathrm{f}})_{S,A,V} = T(\partial S_{\mathrm{f}} / \partial N_{\mathrm{f}})_{E,A,V} ,$$

$$\mu_{\mathrm{v}} = (\partial E_{\mathrm{v}} / \partial N_{\mathrm{v}})_{S,A,V} = T(\partial S_{\mathrm{v}} / \partial N_{\mathrm{v}})_{E,A,V} ,$$

$$\mu_{\mathrm{s}} = (\partial E_{\mathrm{s}} / \partial N_{\mathrm{s}})_{S,A,V} = T(\partial S_{\mathrm{s}} / \partial N_{\mathrm{s}})_{E,A,V} . \qquad (4.2.9\mathrm{a})$$

Thus Eq. (4.2.7) is just the standard condition for thermodynamic equilibrium between two phases, but specifically applied to adsorption:

$$\mu_{\mathrm{f}} = \mu_{\mathrm{s}} . \qquad (4.2.9\mathrm{b})$$

It is in a sense a trivial result, except for two points. First, we emphasize that the derivation does not depend at all upon the sharpness of the division between film and vapor. The equilibrium condition is in fact a special case of the universal condition that the chemical potential of a substance is uniform in thermodynamic equilibrium, even in the presence of arbitrary fields and boundaries. The second point is that it does not require that the substrate be inert, but may in fact participate in the adsorption process in a way that affects its properties rather profoundly.

So far we have shown that the chemical potential is related to certain partial derivatives of the entropy and energy. There are additional connections to other thermodynamic functions, which we now discuss in the context of physical adsorption.

The Helmholtz free energy F, Gibbs free energy G, enthalpy H, and thermodynamic potential Ω for arbitrary systems are

$$F = E - TS, \tag{4.2.10}$$

$$G = F + \sum_\alpha X_\alpha x_\alpha, \tag{4.2.11}$$

$$H = E + \sum_\alpha X_\alpha x_\alpha, \tag{4.2.12}$$

$$\Omega = F - G. \tag{4.2.13}$$

Differentiating each state function and substituting for dE in the form (4.2.8), we obtain

$$dF = -S\,dT - \phi\,dA - P\,dV + \Sigma\,\mu_i\,dN_i, \tag{4.2.14}$$

$$dG = -S\,dT + A\,d\phi + V\,dP + \Sigma\,\mu_i\,dN_i, \tag{4.2.15}$$

where the index i refers to film, vapor, or substrate.

Also,

$$dH = T\,dS + A\,d\phi + V\,dP + \Sigma\,\mu_i\,dN_i, \tag{4.2.16}$$

$$d\Omega = -A\,d\phi - \phi\,dA - V\,dP - P\,dV. \tag{4.2.17}$$

Thus the chemical potentials can be related to the different functions

$$\mu_i = (\partial F/\partial N_i)_{T,A,V,N_j \neq i} = (\partial G/\partial N_i)_{T,\phi,P,N_j \neq i} = (\partial H/\partial N_i)_{S,\phi,P,N_j \neq i}. \tag{4.2.18}$$

Now we can show that the chemical potential is equal to the Gibbs free energy per particle, just as for ordinary bulk matter. According to the division of the composite system into film, vapor, and substrate, we can write Eq. (4.2.15) as three component equations (noting that according to (4.1.20) even P and ϕ can be divided into separate contributions) all of the form

$$dG_i = -S_i\,dT + A\,d\phi_i + V\,dP_i + \mu_i\,dN_i, \tag{4.2.19}$$

and thus the implicit functional form of the Gibbs free energy is, for each of the components,

$$G = G(T, \phi, P, N).$$

But G, just as all of the other functions of state, is an extensive quantity, i.e., proportional to the total number N. Therefore, since all of the other variables T, ϕ, P in its implicit relation are intensive, G must have the functional form

$$G = Ng(T, \phi, P), \tag{4.2.20}$$

and therefore

$$g(T, \phi, P) = G/N = (\partial G/\partial N)_{T,\phi,P}. \tag{4.2.21}$$

But now comparing to Eq. (4.2.18) we see that $g(T, \phi, P)$ is just equal to the chemical potential: in other words, for each component,

$$G_i = N_i\mu_i. \tag{4.2.22}$$

4.3 HOMOGENEITY OF FILM AND VAPOR

Equilibrium in the mixed film–vapor system is characterized by the uniformity of T and μ but not of P, ϕ, or the other intensive variables. Interactions between the adsorbate and the substrate fall off as some function of the distance, and hence P, ϕ, and density vary in the neighborhood of the adsorbing surface. However, the chemical potential of the adsorbate is uniform throughout the equilibrium system; it is in fact the uniformity of μ that makes possible the calculation of local properties.

The proof that μ is uniform in the presence of substrate fields is identical in form with the preceeding proof that $\mu_f = \mu_v$; one could just as well consider an arbitrary division of the adsorbate into two regions having different energies of interaction with the surface. Thus it is not necessary to assume that the film and vapor are homogeneous "phases" although it is often convenient to ignore the variations. The gradients in ϕ can be particularly large due to the proximity of the film to the surface. It is clear that an imagined barrier process such as illustrated in Fig. 4.1 would measure an effective spreading pressure that is not actually a 2D ϕ but some ϕ averaged over the thickness of the edge of the barrier. The actual gradients in real films depend upon the specific interactions between the adsorbate molecules themselves and between adsorbate and substrate. The variations are particularly important in the physics of films of a few molecules thickness. We shall have more to say on this point in Section 4.6, but for the present, we suppress the need for investigating gradients in the film by treating the film as a thin region enclosed between the absorbent surface and Gibbs's fictitious dividing surface. This device creates a sharp division between film and vapor, such that the properties of the film are extensive with surface area and vapor properties are extensive with volume. Now we wish to explore, in terms of simple physical models, to what extent one may actually ignore volume terms in the film and surface terms in the vapor. Here we shall show that the two cross terms are interrelated.

We assume that the vapor behaves as an ideal gas if substrate interactions are absent. The ideality of the vapor is assured in most adsorption work by the typically low pressures; furthermore, deviations from ideality due to molecular interactions or statistical effects are almost always completely masked by adsorption effects. In the absence of substrate fields we have the usual free energy of an ideal classical gas (12):

$$F_v = -N_v kT \ln (e^{3/2} V / N_v \lambda^3), \tag{4.3.1}$$

where $\lambda = h(2\pi mkT)^{-1/2}$ is the thermal de Broglie wavelength. (We assume for simplicity that the molecules have no "internal" degrees of freedom.) From (4.3.1) we obtain

$$S_v = -(\partial F_v / \partial T)_{N_v, V} = N_v k \ln (e^{5/2} V / N_v \lambda^3), \tag{4.3.2}$$

$$P = -(\partial F_v / \partial V)_{T, N_v} = N_v kT / V, \tag{4.3.3}$$

$$\mu_v = (\partial F_v / \partial N_v)_{T, V} = -kT \ln (V / N_v \lambda^3) = -kT \ln (kT / P \lambda^3). \tag{4.3.4}$$

Equations (4.3.1)–(4.3.4) correspond to a hard-wall container and no film present; the dividing surface coincides with the solid boundary of V and there are no long-range substrate fields. If now both film and long-range fields are present (the fields being due to both vapor–substrate and vapor–film interactions) the free energy of the vapor fraction is

$$F_v' = -N_v kT \ln \left\{ (e^{3/2} / N_v \lambda^3) \int_v \exp\left[-\beta u(\mathbf{r})\right] d^3 r \right\}, \tag{4.3.5}$$

where $N_v = N - N_f$ and $u(\mathbf{r})$ is the potential energy of a vapor molecule at $\mathbf{r}$ from the solid wall. $F_v{}'$ is to be compared with the free energy of an ideal gas of N_v molecules in the volume V. This can be done in the style of the theory of dilute imperfect gases (9). We define a "surface second virial coefficient" B_s by

$$B_s \equiv \int_v \{1 - \exp\,[-\beta u(\mathbf{r})]\}\,d^3r. \qquad (4.3.6)$$

Then we can write (4.3.5) in the form

$$\begin{aligned} F_v{}' &= -N_v kT \ln\,[(e^{3/2}V/N_v\lambda^3)(1 - B_s/V)] \\ &= F_v(N_v) - N_v kT \ln\,(1 - B_s/V) \\ &\cong F_v(N_v) + (N_v kT/V)B_s \end{aligned} \qquad (4.3.7)$$

if $B_s/V \ll 1$. $F_v(N_v)$ has the ideal gas form (4.3.1) for N_v molecules. From (4.3.7) we get the perturbed equation of state of the vapor

$$P' = -(\partial F_v{}'/\partial V)_{N_v,T} = (N_v kT/V)(1 + B_s/V) = P(1 + B_s/V), \qquad (4.3.8)$$

where P is the pressure of an ideal gas of N_v molecules. In adsorption work, one is often interested in a particular temperature derivative of pressure known as the "isosteric heat of adsorption" q_{st}. This quantity, which is discussed in more detail later, is defined as

$$q_{st} \equiv kT^2(\partial \ln P/\partial T)_{N_f,A}. \qquad (4.3.9)$$

From (4.3.8) we find that the perturbed heat of adsorption is

$$q_{st}{}' \cong q_{st} + (kT^2/V)(\partial B_s/\partial T)_{N_f,A}. \qquad (4.3.10)$$

Equations (4.3.8) and (4.3.10) show that the departures from ideality caused by vapor–film and vapor–substrate interactions can be reduced to arbitrarily small values by making V arbitrarily large. But this would also have the effect of making all film properties of negligible importance. Thus, it becomes a matter of practical necessity to make some compromise between sensitivity to film effects and a simple equation of state for the vapor.

The corrections can be put on a quantitative basis if we adopt a few simplifying assumptions. We approximate the substrate and film as planar and homogeneous, attracting gas atoms with characteristic energies varying as the inverse cube of the distance (see Chapter 2).

Such a variation is appropriate over intermediate distances ranging from several to perhaps a hundred atomic diameters; for simplicity we shall assume it to hold all the way down to "contact," where repulsion will be approximated as a hard wall. Thus, for an atom at a normal distance z from a substrate, the interaction is represented by

$$u(z) = -u_s/z^3, \qquad z > \sigma, \tag{4.3.11}$$
$$= \infty, \qquad z < \sigma.$$

If the surface is covered by a homogeneous film of thickness d, the total interaction is

$$u(z) = -\frac{(u_s - u_f)}{z^3} - \frac{u_f}{(z-d)^3}, \qquad z > d + \sigma, \tag{4.3.12}$$
$$= \infty, \qquad 0 < z < d + \sigma.$$

We now calculate B_s for this interaction. The repulsive region can be evaluated immediately, but for the attractive interaction we assume a "high temperature" approximation $\beta u(z) \ll 1$, which then allows evaluation of the exponential by the leading term in the power series expansion. This approximation is less stringent than for the adsorbate in general since it refers to the vapor fraction only, and its density is low just because of the smallness of the exponential factor. Thus, carrying out the calculation in the manner described, one obtains

$$B_s \simeq A \int_0^{d+\sigma} dz + A\beta \int_{d+\sigma}^{\infty} u(z)\, dz$$
$$= A(d + \sigma) - \frac{A}{2kT}\left[\frac{u_f}{\sigma^2} + \frac{(u_s - u_f)}{(d+\sigma)^2}\right]. \tag{4.3.13}$$

We are now interested in the relative importance of the surface virial coefficient, as for example, in the extent of its influence on the heat of adsorption. We can put these changes on an interesting comparative basis as follows. The quantity $A(d + \sigma)$ is approximately equal to V_f, the volume occupied by the film, and $A\sigma = V_1$, the volume occupied by a single layer. Then the perturbed heat of adsorption, Eq. (4.3.10), with B_s given by Eq. (4.3.13), can be written

$$q_{st}' - q_{st} = \frac{V_1}{2V}\left[\frac{u_f}{\sigma^3}\right] + \frac{V_f}{2V}\left[\frac{(u_s - U_f)}{(d+\sigma)^3}\right]. \tag{4.3.14}$$

The bracketed factors in both terms are roughly equal to certain heats of adsorption. For example, if there is very little adsorption, $d = 0$ and $V_f = V_1$. Then since $u_s/\sigma^3 \simeq q_{st}$ of the bare substrate, we obtain the fractional correction

$$(q_{st}' - q_{st})/q_{st} \simeq V_1/2V. \qquad (4.3.15)$$

In the case of a thick film, where $d \gg \sigma$,

$$(q_{st}' - q_{st})/q_{st} \simeq V_f/2V, \qquad (4.3.16)$$

where here q_{st} refers to the heat of adsorption (or heat of evaporation) of the bulk liquid. Thus in both examples the fractional correction in the heat of adsorption is comparable to the fractional volume occupied by film. In other words, the deviations from "ideality" of both film and vapor are coupled; the importance of surface terms in the vapor equation of state scales with the volume terms due to the film. Therefore one may reduce both correction terms to extremely small values simply by increasing the vapor volume, but this would cause a loss of ability to detect any of the properties of the film. As a practical matter it is then a compromise between the two trends, and some substrate-associated nonideality of vapor will have to be accepted as a necessary concomitant of sensitivity to film characteristics.

4.4 VAPOR PRESSURE

The equilibrium vapor pressure is determined by the equations of state of film and vapor and the uniformity of chemical potential. If the condition for overall ideality of vapor is satisfied then the equation for P is given by equating μ_v in Eq. (4.3.4) to μ_f, leading to

$$P = (2\pi m/h^2)^{3/2}(kT)^{5/2} \exp (\mu_f/kT). \qquad (4.4.1)$$

Specific equations for P in terms of the physical parameters of the system require detailed microscopic models of the film. However, in this discussion we refrain from the adoption of particular models and instead explore the general interrelationships among the several thermodynamic quantities.

A model independent connection between film and vapor properties can be obtained at this stage. If the area and volume cross terms can be neglected in the free energies of film and vapor, then the differential Gibbs functions are, according to Eq. (4.2.19),

$$dG_f = -S_f \, dT + A \, d\phi + \mu_f \, dN_f, \qquad (4.4.2)$$

$$dG_v = -S_v \, dT + V \, dP + \mu_v \, dN_v. \qquad (4.4.3)$$

With the relations (4.2.22) $G_f = N_f \mu_f$ and $G_v = N_v \mu_v$ in differential form, we obtain

$$d\mu_f = -s_f \, dT + a \, d\phi, \qquad (4.4.4)$$

$$d\mu_v = -s_v \, dT + v \, dP, \qquad (4.4.5)$$

where $s_f = S_f/N_f$, $a = A/N_f$, etc. Equation (4.4.5) is the ordinary Gibbs–Duhem relation for simple bulk matter; Eq. (4.4.4) is sometimes referred to as the Gibbs–Duhem relation for a film. If we now consider an isothermal change in the equilibrium system, i.e., equating (4.4.4) and (4.4.5) and setting $dT = 0$,

$$a \, d\phi = v \, dP. \qquad (4.4.6)$$

Equation (4.4.6) provides, in principle, a means for deducing film properties from the behavior of its equilibrium vapor. If the molecular area a is independently known as a function of vapor pressure then ϕ can be obtained by the integral

$$\phi(P') = \int_0^{P'} (v/a) \, dP, \qquad (4.4.7)$$

where it is assumed that $\phi(0) = 0$. If the vapor behaves as an ideal gas, $v = kT/P$, and we obtain

$$\phi(P') = (kT/A) \int_0^{P'} N_f(P) \, d(\ln P). \qquad (4.4.8)$$

Equation (4.4.7) is known as "Gibbs's adsorption theorem," and (4.4.8) is sometimes called "the Gibbs isotherm." It is a most simple and direct relation between film and vapor quantities, but we note that just as for most of the explicit thermodynamic connections between the "phases," it neglects cross terms and assumes that the vapor is an ideal gas.

If the condition for overall ideality of the vapor is satisfied, this does not automatically imply that there are no gradients in the vapor; it merely means that the inhomogeneous region is quite small. In fact, an argument based upon local ideality, explicitly taking account of spatial variations, leads to a useful gauge of film thickness. Suppose for simplicity that the intrinsic thermodynamic state of the film is the same as that of bulk adsorbate at the same temperature, i.e., that the film is a thin slice of the bulk liquid or solid. In this case the chemical potential of a vapor atom above a film of some thickness d on a substrate differs from the chemical potential of saturated vapor (i.e., in equilibrium with the bulk phase) only due to the energy change caused by the presence of the substrate instead of the deeper regions of the bulk. Thus, if $\mu_{v0}(T)$ is the chemical potential of saturated vapor in equilibrium with the bulk phase, the chemical potential at the surface of a film of thickness d is

$$\mu_v(T, d) = \mu_{v0}(T) + \Delta u(d), \qquad (4.4.9)$$

where $\Delta u(d)$ the change in energy caused by the substitution of substrate for bulk adsorbate below the depth d. If the vapor is locally ideal its chemical potential is in the form given by (4.3.4). Then we obtain

$$\ln (P/P_0) = \beta \, \Delta u(d), \qquad (4.4.10)$$

where P is the vapor pressure of the film and P_0 is the saturated vapor pressure of the bulk. If the interactions are of the type given by (4.3.11) and (4.3.12), then

$$P(T, d) = P_0(T) \exp \left[(u_f - u_s)/kT(d + \sigma)^3 \right]. \qquad (4.4.11)$$

For thick films, $d \gg \sigma$, the exponential factor varies approximately as d^{-3}. The range exponent may have other values (as for example, at long range, where retardation effects tend to produce a d^{-4} variation). In its most general form the exponent varies as d^{-n}. With slight variations the simpler power law version was proposed independently by Frenkel (14), Halsey (6), and Hill (15). We emphasize that Eq. (4.4.11) and the FHH isotherm are based upon the assumption that the film is a uniform slice of the bulk phase. This condition must be approached asymptotically as d increases, but it can only be an approximate relation for films of a few layers' thickness.

4.5 HEAT CAPACITY AND ENTROPY

Direct measurement of the total heat capacity of the system yields contributions from film, vapor, and substrate. For fixed N, F, A the total heat capacity

$$T(\partial S/\partial T)_{N,V,A} = T[(\partial/\partial T)(S_f + S_v + S_s)]_{N,V,A}. \qquad (4.5.1)$$

The substrate contribution can be measured in the absence of adsorbate and subtracted from the total. We are then left with the adsorbate contribution which we denote as C_{eq}. This "equilibrium heat capacity" C_{eq} is composed of three terms: parts due to fixed quantities of film and vapor, and a contribution due to film–vapor conversion along the equilibrium curve. Thus, we can write

$$(\partial S_f/\partial T)_{N,A} = (\partial S_f/\partial T)_{N_f,A} + (\partial S_f/\partial N_f)_{T,A}(dN_f/dT)_{eq}, \qquad (4.5.2)$$

and a corresponding expression for the vapor. We define C_f, C_v as the heat capacities at constant N_f and N_v,

$$C_f \equiv T(\partial S_f/\partial T)_{N_f,A}, \qquad C_v \equiv T(\partial S_v/\partial T)_{N_v,V}. \qquad (4.5.3)$$

Then the equilibrium heat capacity for the combined film and vapor is

$$\begin{aligned}
C_{eq} &= T\{(\partial/\partial T)(S_f + S_v)\}_{N,V,A} \\
&= C_v + C_f + T[(\partial S_v/\partial N_v)_{T,V} - (\partial S_f/\partial N_f)_{T,A}](dN_v/dT)_{eq}.
\end{aligned} \qquad (4.5.4)$$

The difference in partial entropies (the term in brackets) in (4.5.4) is related to the heat of adsorption introduced earlier. Using the expression (4.3.2) for the entropy of the vapor, we get

$$(\partial S_v/\partial N_v)_{T,V} = S_v/N_v - k. \qquad (4.5.5)$$

For the partial entropy of the vapor, we write

$$(\partial S_f/\partial N_f)_{T,A} = -\partial^2 F_f/\partial N_f\,\partial T = -(\partial \mu_f/\partial T)_{N_f,A}. \qquad (4.5.6)$$

Now, with the defining equation (4.3.9) for q_{st} in terms of pressure, and substituting the general relation (4.4.1) for P, we have

$$q_{st} = kT^2(\partial \ln P/\partial T)_{N_f,A} = \tfrac{5}{2}kT - \mu_f + T(\partial \mu_f/\partial T)_{N_f,A}. \qquad (4.5.7)$$

With $\mu_f = \mu_v$ and again using (4.3.2) for S_v, the first two terms on the right-hand side of (4.5.7) can be written

$$\tfrac{5}{2}kT - \mu_f = \tfrac{5}{2}kT - \mu_v = TS_v/N_v. \qquad (4.5.8)$$

Combining (4.5.6) and (4.5.8) into (4.5.7), we get a general relation between q_{st} and the vapor and film entropies:

$$q_{st} = T[(S_v/N_v) - (\partial S_f/\partial N_f)_{T,A}]. \qquad (4.5.9)$$

Equation (4.5.9) provides a thermodynamic basis for determining the partial entropy of a film from vapor pressure measurements, based of course on the assumption of ideality.

The same quantity q_{st} also enters into the heat capacity of the mixed film–vapor system. Substituting (4.5.9) and (4.5.4) in Eq. (4.5.4) for the heat capacity C_s of the substrate, we obtain

$$C_{eq} = C_v + C_f + (q_{st} - kT)(dN_v/dT)_{eq}. \qquad (4.5.10)$$

We now explore the interrelationships among heat capacity, vapor pressure, and heat of adsorption in order to extract some expressions of practical usefulness.

In Eq. (4.5.10) the conversion term $(dN_v/dT)_{eq}$ can itself be related to q_{st}. Differentiating the ideal gas law (4.3.3), we have

$$\left(\frac{dN_v}{dT}\right)_{eq} = \frac{d}{dT}\left(\frac{PV}{kT}\right) = -\frac{PV}{kT^2} + \frac{PV}{kT}\left(\frac{d\ln P}{dT}\right)_{eq}. \qquad (4.5.11)$$

The pressure derivative is the sum of two terms

$$\left(\frac{d\ln P}{dT}\right)_{eq} = \left(\frac{\partial \ln P}{\partial T}\right)_{N_f,A} + \left(\frac{\partial \ln P}{\partial N_f}\right)_{T,A}\left(\frac{dN_f}{dT}\right)_{eq}. \qquad (4.5.12)$$

With the definition for q_{st} and $dN_f = -dN_v$, Eqs. (4.5.11) and (4.5.12) lead to

$$\left(\frac{dN_v}{dT}\right)_{eq} = \frac{PV}{k^2T^3}(q_{st} - kT)\left[1 + \frac{PV}{kT}\left(\frac{\partial \ln P}{\partial N_f}\right)_{T,A}\right]^{-1}. \qquad (4.5.13)$$

Substituting (4.5.13) for (dN_v/dT) in (4.5.10) and using the ideal gas heat capacity $C_v = \tfrac{3}{2}PV/T$, one obtains (16)

$$C_{eq} = C_f + \frac{PV}{T}\left\{\frac{3}{2} + \left(\frac{q_{st}}{kT} - 1\right)^2\left[1 + \frac{PV}{kT}\left(\frac{\partial \ln P}{\partial N_f}\right)\right]^{-1}\right\}. \qquad (4.5.14)$$

Therefore (4.5.14) shows that C_f can be extracted from measurements of the total heat capacity if one has ancillary data from vapor pressure isotherms and isosteres. The contribution due to desorption $C_{des} = C_{eq} - C_f$ becomes relatively unimportant at low temperatures; there is always a range in which C_{des} may be ignored. This is because P varies approximately as $\exp(-q_{st}/kT)$, while the temperature dependence of C_f is invariably less rapid.

If C_{des} is not negligible then it is possible to deduce P from heat capacity measurements alone. This can be done by comparing total heat capacities of systems having different values of V but identical in all other respects, which can be done by changing the part of the experimental volume that is at room temperature. The theoretical connection between P and C_{des} is obtained by equating the expressions for $(\partial q_{st}/\partial T)_{N_f}$ that can be obtained from Eqs. (4.5.14) and (4.3.9). This yields an exact expression which can be solved numerically for P. When $q_{st}/kT \gtrsim 10$, the solution to within a few percent is (16)

$$P = (TC_{des}/V)[2 + T(\partial \ln C_{des}/\partial T)_{N_f}]^{-2}. \qquad (4.5.15)$$

If the total heat capacity is measured down to such low temperatures that one can make a reliable extrapolation to $T = 0$, and if a single value of P is known at some point within the measured range of $C_{eq}(N, T)$, then one can calculate the absolute entropy $S_f(N_f, T)$ of the film. The total entropy of the adsorbate is

$$S(N, T) = \int_0^T C_{eq}(N, T)(dT/T) = S_f(N_f, T) + S_v(N_v, T). \qquad (4.5.16)$$

Using Eqs. (4.3.2) and (4.3.3) for the vapor, we obtain

$$S_f(N_f, T) = \int_0^T C_{eq}(N, T)\frac{dT}{T} - \frac{PV}{T} \ln\left[\frac{e^{5/2}kT}{P}\left(\frac{2\pi mkT}{h^2}\right)^{3/2}\right], \qquad (4.5.17)$$

where $N_f = N - PV/kT$. Equation (4.5.17) is based on the assumption that the vapor is an ideal classical gas but it does not depend on any specific model of the film. It does not require that the desorption contribution to C_{eq} be separately measured or calculated: the effects of desorption are automatically included in $S_v(N_v, T)$. If Eq. (4.5.15) is combined with (4.5.17), then the absolute entropy of a film could be calculated from purely thermal measurements. This would be subject, of course, to the condition that the vapor pressure is sufficiently low for (4.5.15) to be satisfactory. In the most fortunate circumstances P will be so small that the entire second term on the right-hand side of Eq. (4.5.17) can be neglected, in which case S_f will be obtained simply by integrating C_{eq}. The major uncertainty in this evaluation is then likely to be in the extrapolation of C_{eq} from the lowest experimental temperature to $T = 0$, and this will depend upon both instrumental factors and questions concerning the detailed theoretical model of the film.

An alternative evaluation of the absolute entropy of the film can be made by vapor pressure measurements alone. Since

$$S_f(N_f, T) = \int_0^{N_f} (\partial S_f / \partial N_f)_{T,A} \, dN_f, \qquad (4.5.18)$$

using Eq. (4.5.9) for the partial entropy of the film and the ideal gas relations (4.3.2–4.3.4), we obtain

$$S_f(N_f, T) = \tfrac{5}{2} N_f k - (1/T) \int_0^{N_f} (\mu_v + q_{st}) \, dN_f. \qquad (4.5.19)$$

All of the quantities on the right-hand side of (4.5.19) can be determined through vapor pressure measurements. This method of determining the absolute entropy of films, while quite direct according to the formal equations, is generally less reliable than by thermal measurements. This is because the terms μ_v and q_{st} are generally comparable in magnitude and of opposite sign, and their difference can in real cases amount to less than 1% of their absolute magnitudes.

4.6   SURFACE PHASES

If the gradients caused by the substrate fields are ignored and we consider the adsorbate to be composed simply of film and vapor, then there can be as many as four coexisting phases of the adsorbate. This follows from the fact that the system would be described by four pairs of conjugate variables: (T, S); (P, V); (ϕ, A); (μ, N). Such a "quadruple point" would be the analogue to the triple point of conventional bulk systems, and could consist of three surface phases (e.g., "2D solid, liquid, and vapor" phases) in equilibrium with the 3D vapor. If the adsorption system consists of only three phases, then these can coexist along an equilibrium line $P = P(T, \phi)$, and two phases can coexist over a finite area of the P, T, ϕ surface.

But since substrate fields are in fact present in both the film and vapor fractions the situation can be much more complicated. The substrate attraction causes gradients in the adsorbate, particularly in those layers of film closest to the substrate, and these may be so great that adjacent layers can behave with thermodynamic individuality. Properties such as surface density of mass and entropy in each layer may be so different as to lead to quite distinct phase transitions. Such behavior is experimentally observed in a number of systems described in Chapters 6 and 7. In helium films on graphite, for instance, a three-layer film at sufficiently low temperature can consist of a dense solid first layer, a less dense solid second layer, and a liquid or vapor top layer. The two solid layers, by reason of their different densities and environments, have different structural arrangements (there being no limit to the variety of superlattice structures; see Chapter 8). By the criterion of distinct symmetries (12), under these conditions each structure constitutes a distinct phase. The three-layer helium film has markedly different melting transitions for each layer. At relatively high temperature, the three layers lose their individuality and become a single inhomogeneous phase, but at low temperature, the layer structure is thermodynamically discrete.

One is then led to consider progressively thicker films, inquiring whether they might also exhibit layer behavior, or whether there is some critical thickness beyond which there can be only a single inhomogeneous phase. Apart from the fact that the differences between layers become progressively more delicate further from the substrate, and that therefore the temperature range for layer individuality is correspondingly lowered, there seems no theoretical limit to the thickness of layer structure. Experimental limits are imposed by the temperatures, instrumental resolution, and substrate uniformity.

This analysis extends the arguments originally given by Gibbs concerning the "dividing surface" as a device for distinguishing between film and vapor. In the following discussion we consider the possible

changes that can occur in a single layer, bearing in mind that simultaneous changes in other layers may also occur.

In the event that there are two or more surface phases coexisting
with each other in a layer and each with the vapor at a single set of
variables (P, T, ϕ), then a number of distinctive features will be exhibited by the thermal properties of the combined systems. These
features, as will be shown, depend primarily on the fact of phase
coexistence and not on the detailed nature of the phases.

The fundamental attribute of the region of surface phase coexistence
is that the chemical potential is independent of N, being "clamped" by
the fact that there is a single equilibrium line $P = P(T, \phi)$ for the
system. The chemical potentials of the two surface phases (1, 2) and
the vapor are all equal:

$$\mu_1(\phi, T) = \mu_2(\phi, T) = \mu_v(P, T) \qquad (4.6.1)$$

independent of N all along the coexistence line. Since the vapor pressure
is directly related to the chemical potential [Eq. (4.4.1)], then P also
must be independent of N. Thus, the region of two-phase equilibrium
on the surface must show up as a vertical portion of a vapor pressure
isotherm, as illustrated schematically in Fig. 4.2. At the same time, the
isosteric heat of adsorption q_{st} will be independent of N.

The equilibrium heat capacity will also have a distinctive signature
in a two-phase region. If we consider the total differential Gibbs free
energy of the mixed system, for a constant quantity of adsorbate, then
Eq. (4.2.15) yields

$$dG = -S\,dT + A\,d\phi + V\,dP. \qquad (4.6.2)$$

Now, consider the first and second temperature derivatives of G at
constant area and volume (the typical conditions of a measurement of
heat capacity); differentiating (4.6.2), we obtain

$$\left(\frac{dG}{dT}\right)_{eq} = N\left(\frac{d\mu}{dT}\right)_{eq} = -S + A\left(\frac{d\phi}{dT}\right)_{eq} + V\left(\frac{dP}{dT}\right)_{eq}, \qquad (4.6.3)$$

$$\left(\frac{d^2G}{dT^2}\right)_{eq} = N\left(\frac{d^2\mu}{dT^2}\right)_{eq} = -\frac{C_{eq}}{T} + A\left(\frac{d^2\phi}{dT^2}\right)_{eq} + V\left(\frac{d^2P}{dT^2}\right)_{eq}. \qquad (4.6.4)$$

It is the fact that there is a single equilibrium line $P(T, \phi)$ that makes

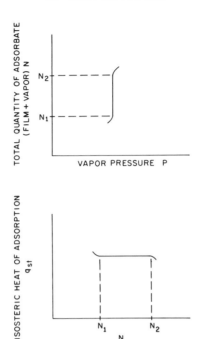

FIG. 4.2 Characteristic signatures of the vapor pressure isotherm and isosteric heat of adsorption during two-phase equilibrium in a film.

the several equilibrium derivatives independent of N and hence universal functions of T alone throughout the phase coexistence region. Rearranging Eq. (4.6.4), we obtain the general condition on the total heat capacity (17)

$$C_{eq} = Nf(T) + g(T). \qquad (4.6.5)$$

Thus the "signature" of C_{eq} during phase equilibrium has the form illustrated in Fig. 4.3. It must be noted, however, that Eq. (4.6.5) is a necessary but not sufficient concomitant of phase coexistence; it is possible to have homogeneous surface phases with the same functional dependence as (4.6.5).

There are quite distinctive discontinuities at the boundaries of two-phase regions, i.e., where one of the phases just appears or disappears. To see this all one has to do is to reexamine Eq. (4.5.4), which was

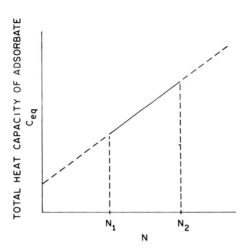

FIG. 4.3   The heat capacity of the adsorption system varies linearly with total quantity of adsorbate when the film is undergoing two-phase equilibrium.

derived earlier in the discussion of film–vapor equilibrium. If now there are two surface phases there will be two parts to C_f and another factor involving a difference of partial entropies, i.e., a term

$$T \left[\left(\frac{\partial S_{f_1}}{\partial N_{f_1}} \right)_{T,A_1} - \left(\frac{\partial S_{f_2}}{\partial N_{f_2}} \right)_{T,A_2} \right] \left(\frac{dN_{f_1}}{dT} \right)_{eq}. \qquad (4.6.6)$$

This term represents the progressive conversion from phase 2 to phase 1 as T changes. It is present only during phase coexistence; thus it appears or disappears abruptly with the first and last traces of a second phase. To give this discussion a more definite context, we imagine two different examples: surface phase evaporation and surface phase melting.

For evaporation, we imagine that the film is composed of condensed (liquid or solid) patches in equilibrium with a low density "2D vapor." As the combined system is warmed the 2D vapor pressure ϕ will in general increase, and the density of the 2D vapor phase will increase accordingly. As the evaporation process continues there will be a term of the form given in (4.6.6) contributing to the heat capacity. Finally, a temperature will be reached at which the last trace of condensed phase evaporates; when this happens, both the conversion contribution and C_f (condensed) terms disappear abruptly; the total heat capacity has the shape illustrated in Fig. 4.4. The proof of the existence of the discontinuity does not require the assumption of any definite model of the condensed phase or of the vapor. However, fairly general considerations predict that the overall shape of the conversion term will contain an exponential factor of the activation energy type, i.e., of the form

$$\phi \sim \exp\left(-\text{const}/T\right).$$

This is readily obtained by first constructing the "surface Clausius–Clapeyron equation"; equating relations of the form (4.4.5) for two surface phases, we have

$$(d\phi/dT)_{eq} = (s_{f_1} - s_{f_2})/(a_{f_1} - a_{f_2}). \tag{4.6.7}$$

Then, if one follows the same sort of approximation as in the elementary treatment of evaporation, i.e., assuming that the vapor is an ideal gas of much lower density than the condensed phase, and that the latent heat $q_{vap} = T\,\Delta s$ is a constant, then Eq. (4.6.7) leads to the 2D vapor pressure equation of the form given above, i.e.,

$$\phi = \phi_0 e^{-q/kT}. \tag{4.6.8}$$

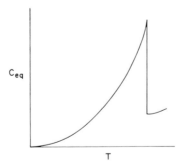

FIG. 4.4 Schematic illustration of the temperature dependence of the equilibrium heat capacity for a film composed of 2D condensed patches in equilibrium with 2D vapor. The discontinuous drop occurs when the last trace of condensed phase evaporates.

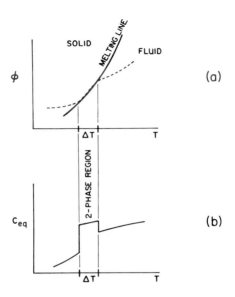

FIG. 4.5   Film having liquid and solid phases might have a melting line such as illustrated in (a). If a sample is heated under constant area conditions its $\phi(T)$ trajectory could resemble the dashed line in (a). Then the total heat capacity of adsorbate would have the form shown in (b), with finite discontinuities due to the first appearance of "liquid" and the final disappearance of "solid." The discontinuities are predicated upon the existence of a first-order melting transition in the solid phase, which depends upon special conditions in the particular adsorption system (see Chapter 7).

The shape of the heat capacity around a region of surface phase "melting" has a still different shape. We imagine an initial situation in which a surface is completely covered with some "solid" film phase at a low temperature and then gradually heated. In the region of pure solid phase there is no conversion term; such a contribution will appear at some definite temperature together with the initial appearance of "liquid" film. Thus C_{eq} takes a discontinuous jump, and in the coexistence region there are three terms due to solid, liquid, and conversion. As T rises, the surface pressure ϕ (and also P) will change, and the mixed system will move along the solid–liquid equilibrium curve as illustrated in Fig. 4.5, i.e., with a steadily changing "melting" temperature. Finally, with the melting of the last bit of solid, the heat capacity drops discontinuously down to a single term, the homogene-

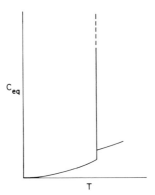

FIG. 4.6   At the "quadruple point" of the adsorbate (the triple point of the monolayer) a film melts at constant ϕ. Under these conditions the total heat capacity becomes infinite, just as for C_P in a bulk system at its triple point. The singularity shown here presumes first-order melting in the solid phase as in Fig. 4.5.

ous liquid phase. [All of this paragraph is predicated, of course, on the supposition that there can indeed be first-order melting phase changes in a film. This is not necessarily the case, however; questions involving the order of the phase change in melting of films are discussed in Chapter 7.]

The process just considered is at constant area. It is also possible to imagine a process at constant ϕ. In this case, one is considering melting at a quadruple point (i.e., a monolayer triple point), i.e., melting of solid patches which are in equilibrium with 2D vapor. Then, if the melting process is a first-order phase transition, there must be a finite amount of heat added while the system remains at a fixed temperature: C_ϕ is infinite, just as C_P is in a first-order process in a 3D system (Fig. 4.6). But this is again predicated on the assumption that melting of the film is indeed a first-order process. As mentioned above, this may or may not be the case in a particular adsorption system.

A few more properties such as the course of the vapor pressure isotherms and isosteres can be worked out for the various cases. For more detailed descriptions it is necessary to assume specific microscopic models of the surface phases. This can be readily done within the framework of the general theory, using relatively standard techniques and several of the relations given in the previous section. Numerous models are explored in subsequent chapters.

4.7 HEATS OF ADSORPTION

Adsorption is an exothermic process. The energetics of the process is described in terms of a heat of adsorption. In the limiting case $T = 0$ and zero coverage, the heat of adsorption is equal to the binding energy of atoms to the substrate. But at finite temperature and coverage the heat of adsorption also reflects the states of both film and vapor: it is the heat that must be added to change the states of particles from one "phase" to the other. Therefore, assuming that the vapor state is known, and indeed we assume it to be an ideal gas in most practical cases, the heat of adsorption can serve as a probe of the state of the film. However, since for thin films, the attraction to the substrate is usually the dominant term in the heat of adsorption, such measurements are primarily useful for determining film thickness changes and surface heterogeneity, not for the generally weaker effects of "lateral" interactions. These are more readily examined by measurements of the heat capacity. In this section we derive several relationships involving heats of adsorption measured under a variety of experimental conditions.

Under isothermal conditions the heat of adsorption is a differential quantity

$$q \equiv (\partial Q_R / \partial N_f)_T , \ldots , \qquad (4.7.1)$$

where dQ_R is the heat given to a thermal reservoir when dN_f is transformed from vapor to film. Different heats of adsorption refer to specific processes in which particular sets of thermodynamic variables are held fixed. The "differential heat of adsorption" q_d is defined as

$$q_d \equiv (\partial Q_R / \partial N_f)_{T,V,A} . \qquad (4.7.2)$$

q_d is simply related to q_{st}. Since $dQ_R = -dQ = -T \, dS$,

$$q_d = -T(\partial S / \partial N_f)_{T,V,A} = -T[(\partial S_f / \partial N_f)_{T,A} - (\partial S_v / \partial N_v)_{T,V}], \qquad (4.7.3)$$

where we have used $dN_v = -dN_f$. Substituting Eq. (4.5.5) for the partial entropy of the (ideal) vapor in (4.7.3) and comparing with

(4.5.9), we have

$$q_{\mathrm{d}} = q_{\mathrm{st}} - kT. \tag{4.7.4}$$

The isosteric heat of adsorption is equal to the differential heat that would be obtained in a constant pressure process. Thus, if T, P, A are held constant,

$$q_{T,P,A} = -T(\partial S/\partial N_{\mathrm{f}})_{T,P,A} = -T[(\partial S_{\mathrm{f}}/\partial N_{\mathrm{f}})_{T,A} - (\partial S_{\mathrm{v}}/\partial N_{\mathrm{v}})_{T,P}]. \tag{4.7.5}$$

Substituting (4.3.3) into (4.3.2), we can express the vapor entropy in terms of T, P, N_{v}, from which we obtain

$$(\partial S_{\mathrm{v}}/\partial N_{\mathrm{v}})_{T,P} = S_{\mathrm{v}}/N_{\mathrm{v}}. \tag{4.7.6}$$

Substituting (4.7.6) into (4.7.5) and comparing with (4.5.9), we have

$$q_{T,P,A} = T[(S_{\mathrm{v}}/N_{\mathrm{v}}) - (\partial S_{\mathrm{f}}/\partial N_{\mathrm{f}})_{T,A}] = q_{\mathrm{st}}. \tag{4.7.7}$$

Neither q_{d} nor q_{st} correspond to the usual experimental processes in measuring heats of adsorption. In typical experiments a finite quantity of gas is added to the adsorption cell from an "external" bulb. After equilibration it is found that the pressure is lower than if all of the increment had remained in the (ideal) gas phase; this is identified as ΔN_{f}. If $\Delta N_{\mathrm{f}}/N_{\mathrm{f}}$ is small, then the finite process approximates the differential q. The actual process is irreversible; in order to relate it to the equilibrium properties of the system, the measured heat must be replaced by an idealized reversible process. The external bulb is replaced by a piston by which the volume V is changed by a differential amount. In this process N and T remain constant while a differential quantity of vapor is converted reversibly to the film phase. The difference between this process and the constant volume process which yields q_{d} is that there is an additional contribution from the heat of compression $(2, 7)$. For the piston process the differential heat is termed "the isothermal calorimetric heat of adsorption," q_{th}. According to the constants of this process,

$$q_{\mathrm{th}} \equiv (\partial Q_{\mathrm{R}}/\partial N_{\mathrm{f}})_{T,A} = -T(\partial S/\partial N_{\mathrm{f}})_{T,A}. \tag{4.7.8}$$

Expressing S in terms of S_{f} and S_{v} as before, we obtain a term $(\partial S_{\mathrm{v}}/\partial N_{\mathrm{f}})_{T,A}$. S_{v} is a function of A only indirectly, through the equation of state of the combined film–vapor system. The cross-connection can be obtained explicitly through (4.3.2) and (4.3.3). Thus from (4.3.2), we have

$$\left(\frac{\partial S_v}{\partial N_v}\right)_{T,A} = \frac{S_v}{N_v} - k + \frac{N_v k}{V}\left(\frac{\partial V}{\partial N_v}\right)_{T,A}. \qquad (4.7.9)$$

Using (4.3.3) and $dN_v = -dN_f$, we obtain

$$\left(\frac{\partial V}{\partial N_v}\right)_{T,A} = \left[\frac{\partial}{\partial N_v}\left(\frac{N_v kT}{P}\right)\right]_{T,A} = \frac{kT}{P} + \frac{V}{P}\left(\frac{\partial P}{\partial N_f}\right)_{T,A}. \qquad (4.7.10)$$

Substituting (4.6.9) and (4.6.10) into (4.6.8), we obtain

$$q_{th} = T\left[\frac{S_v}{N_v} - \left(\frac{\partial S_f}{\partial N_f}\right)_{T,A}\right] + V\left(\frac{\partial P}{\partial N_f}\right)_{T,A} = q_{st} + V\left(\frac{\partial P}{\partial N_f}\right)_{T,A}. \qquad (4.7.11)$$

The pressure term can be obtained from a vapor pressure isotherm.

Now we consider a reversible adiabatic process in which the thermally isolated system experiences a change in temperature when N_f is changed. In this case we begin with

$$T\,dS = T\,d(S_a + S_f + S_v) = 0, \qquad (4.7.12)$$

where S_a is the entropy of adsorbent and container. Treating the entropies as the implicit functions $S_f(T, A, N_f)$, $S_v(T, P, N_f)$, we have for constant A

$$T\,dS_f = C_f\,dT + T(\partial S_f/\partial N_f)_{T,A}\,dN_f, \qquad (4.7.13)$$

$$T\,dS_v = C_{vP} + T(\partial S_v/\partial N_v)_{T,P}\,dN_v + T(\partial S_v/\partial P)\,dP, \qquad (4.7.14)$$

where C_{vP} is the total heat capacity of vapor at constant N_v, P. The second and third partials of S_v are found from (4.3.2) and (4.3.3), and we obtain for (4.7.14)

$$T\,dS_v = C_{vP}\,dT - (S_v/N_v)\,dN_v - V\,dP. \qquad (4.7.15)$$

Substituting in (4.7.12) and collecting terms, we have

$$(C_a + C_f + C_{vP})\,dT + T[(\partial S_f/\partial N_f)_{T,A} - (S_v/N_v)]\,dN_f - V\,dP = 0,$$

from which we obtain

$$(C_a + C_{vP} + C_f)(\partial T/\partial N_f)_S = q_{st} + V(\partial P/\partial N_f)_S. \quad (4.7.16)$$

The "adiabatic heat of adsorption," q_s, is defined by (7)

$$q_s = (C_{aP} + C_{vP} + C_{fP})(\partial T/\partial N_f)_S, \quad (4.7.17)$$

where the component heat capacities refer to constant P as well as to constant A and numbers of particles. Since C_a, C_f are independent of P, (4.7.16) and (4.7.17) yield

$$q_s = q_{st} + V(\partial P/\partial N_f)_S. \quad (4.7.18)$$

REFERENCES

1. J. W. Gibbs, *Trans. Conn. Acad.* **3**, 108, 343 (1878); "Scientific Papers." Longmans, Green, Boston, Massachusetts, 1906; reprinted by Dover, New York, 1961.
2. A. F. H. Ward, *Proc. Roy. Soc. (London)* **A133**, 506 (1931).
3. M. Polanyi, *Z. Electrochem.* **35**, 431 (1929).
4. E. A. Guggenheim, *Trans. Faraday Soc.* **36**, 397 (1940).
5. T. L. Hill, *J. Chem. Phys.* **17**, 520 (1949).
6. G. D. Halsey, Jr., *J. Chem. Phys.* **16**, 931 (1948).
7. G. L. Kington and J. G. Aston, *J. Amer. Chem. Soc.* **75**, 1924 (1951).
8. D. H. Everett and D. M. Young, *Trans. Faraday Soc.* **48**, 1164 (1952).
9. W. A. Steele and G. D. Halsey, Jr., *J. Chem. Phys.* **23**, 979 (1954).
10. D. M. Young and A. D. Crowell, "Physical Adsorption of Gases." Butterworths, London and Washington, D.C., 1962, references to Chapter 3.
11. R. Fowler and E. A. Guggenheim, "Statistical Thermodynamics." Cambridge Univ. Press, London and New York, 1949.
12. L. Landau and E. M. Lifshitz, "Statistical Physics." Pergamon, Oxford, 1958.
13. See for example, R. J. Tykodi, *J. Chem. Phys.* **22**, 1647 (1954); M. Schick and C. E. Campbell, *Phys. Rev.* **A2**, 1591 (1970).
14. J. Frenkel, "Kinetic Theory of Liquids." Oxford Univ. Press, London and New York, 1949.
15. T. L. Hill, *J. Chem. Phys.* **17**, 590 (1949).
16. J. G. Dash, R. E. Peierls, and G. A. Stewart, *Phys. Rev.* **A2**, 932 (1970).
17. G. A. Stewart and J. G. Dash, *Phys. Rev.* **A2**, 918 (1970).

5. Thermodynamics of Noninteracting Monolayers

We begin with two very simple models; the lattice gas model originally proposed by Langmuir (1, 2) and the ideal 2D gas (2, 3). Their thermal properties are readily derived by following the procedures of general statistical physics, and therefore they serve to illustrate some general techniques useful for more realistic theories. The Langmuir and 2D gas models are also interesting because they lie at opposite extremes of a mobility scale. The atoms of an ideal 2D gas are completely free to move about on the surface, and hence this model is sometimes referred to as "completely mobile." In contrast, the Langmuir model assumes that adsorbed atoms are restricted to definite adsorption sites and thus is "completely localized." (The Langmuir model is the simplest, "noninteracting" example of a wider class of theoretical lattice gases, having various ranges and types of interactions between atoms adsorbed on different sites, discussed in some detail in Chapter 8.) Faced with these two models, one may ask whether there are any theories involving intermediate mobilities, such that changes in a few parameters would cause an approach to either extreme. The question was raised many years ago, but apparently no theoretical attempts were published before Hill presented a semiclassical approximation (4) for what has come to be called "the localized-mobile transition." More recently it has been shown that intermediate mobility, for noninteracting and weakly interacting particles, can be

93

treated completely quantum mechanically in a straightforward manner in terms of surface band states (5). The quantum band theory indicates that there can be two regions of mobility; a high temperature regime of "thermal mobility" and a low temperature regime of "quantum mobility." These points lead to an inquiry concerning the operational meaning of mobility in different experiments. Examples can be found for which a film may appear "mobile" with respect to certain properties but simultaneously "localized" with respect to others. Nevertheless, it is possible to define criteria for mobility with regard to any particular property by a detailed comparison with limiting cases.

The question of mobility is actually much more difficult than it may seem from the simple theoretical models discussed in this chapter. When interactions between the adsorbed atoms are important the nature of the "localized-mobile transition" can change from a single-particle property to a collective transition of the entire film. An analogous transition exists in electronic solids and this "metal-insulator transition" is a topic of considerable interest. We postpone its discussion until Chapter 8.

5.1   LANGMUIR MONOLAYERS†

In this model the substrate has some fixed number (N_s) of identical sites, each capable of adsorbing one gas atom. An adatom has some definite energy ϵ_0 binding it to its site and no other "internal" energy spectrum.

The thermodynamical properties of the Langmuir model are readily obtained by conventional statistical mechanics together with some special details described in Chapter 4. Since the adsorbed atoms are assumed to be completely localized on separate distinct sites, the film obeys Boltzmann statistics. The partition function Z_f for N_f adsorbed atoms is readily evaluated as follows.

$$Z_f = \sum_i \exp\left(-\beta E_i\right) = Q(N_f, N_s) \exp\left[-\beta E(N_f)\right], \quad (5.1.1)$$

† Langmuir (1); Fowler and Guggenheim (2).

where

$$Q(N_f, N_s) = N_s!/N_f!(N_s - N_f)!$$ (5.1.2)

is the configurational degeneracy factor corresponding to the number of distinct arrangements for N_f atoms on N_s sites. With Stirling's approximation for the factors in (5.1.2), we obtain

$$F_f = -kT \ln Z_f$$
$$= -N_f\epsilon_0 + N_s kT[x \ln x + (1 - x) \ln (1 - x)], \quad (5.1.3)$$

where $x \equiv N_f/N_s$.

The vapor pressure P is calculated by equating chemical potentials of film and vapor. Using the ideal gas expression

$$\mu_v = -kT \ln [(kT/P)(2\pi mkT/h^2)^{3/2}]$$ (5.1.4)

for the vapor, and evaluating $\mu_f = (\partial F_f/\partial N_f)_{T,N_s}$, the resulting vapor pressure relation is

$$P = (2\pi m/h^2)^{3/2}(kT)^{5/2}[x/(1 - x)] \exp (-\epsilon_0/kT).$$ (5.1.5)

The coverage dependent factor of Eq. (5.1.5) is known as "the Langmuir isotherm," which is linear at low coverage and diverges as x approaches unity as illustrated in Fig. 5.1. The trend toward linear behavior at small x is a general feature of all models, arising from the unimportance of adatom–adatom interactions at sufficiently low coverage. Although the Langmuir model is usually classed as a "noninteractive" monolayer, this actually refers only to atoms on different

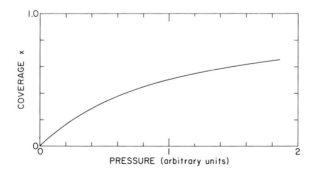

FIG. 5.1   Vapor pressure isotherm of a Langmuir monolayer.

sites. The physical basis of the "site exclusion" which limits occupation number to a maximum of 1 are the short-range repulsive interactions between the adatoms. These interactions cause the pressure to
diverge as the monolayer becomes "completed." The rise in pressure
associated with completion has nothing to do with the assumption of
localization. A very similar coverage dependence is obtained from a
comparably simple model of a mobile film, a 2D hard-disk classical gas
(see Chapter 6).

From (5.1.5) one can obtain the isosteric heat of adsorption q_{st} :

$$q_{st} = kT^2(\partial \ln P/\partial T)_x = \epsilon_0 + \tfrac{5}{2}kT. \qquad (5.1.6)$$

This result is particularly interesting compared to (5.2.11) for the 2D
gas, for it shows how insensitive the heat of adsorption is to mobility.
The heat capacity, on the other hand, is extremely sensitive. For the
Langmuir model, the heat capacity is zero at all temperatures. This is
seen directly from the fact that the entropy is independent of T :

$$S_f = -(\partial F_f/\partial T)_{N_{s,x}} = -N_s k[x \ln x + (1 - x) \ln (1 - x)]. \qquad (5.1.7)$$

The constancy of S_f follows from the complete energy degeneracy of
the various configurations and the lack of any other excitations. The
film entropy arises from the "mixing" of occupied and unoccupied
sites, and just as in the familiar treatments of mixtures of different
species, the mixing entropy does not vanish at $T = 0$. Additional discussion of these points is given in Section 5.3.2.

5.2   IDEAL TWO-DIMENSIONAL GASES†

5.2.1   Boltzmann Approximation

We assume a uniform smooth attracting substrate of area A on
which N_f particles are adsorbed. Each particle has binding energy ϵ_0
with respect to surface–normal motions but can translate freely in
the x, y surface–parallel directions. The energy function $E(p, r)$ of the

† Fowler and Guggenheim (2), Band (3).

film is

$$E(p, r) = -N_f \epsilon_0 + \sum_i^{N_f} (p_i{}^2/2m), \qquad (5.2.1)$$

where $p_i{}^2 = p_{ix}{}^2 + p_{iy}{}^2$; $0 < p_{ix,y} < \infty$. At sufficiently high temperatures and low densities the Boltzmann approximation can be used The condition for its validity is

$$n\lambda^2 \ll 1, \qquad (5.2.2)$$

where $n = N_f/A$; $\lambda = h(2\pi mkT)^{-1/2}$ We call $n\lambda^2$ the "2D degeneracy factor" by analogy with the usual factor encountered in bulk systems. In this regime we can obtain thermal properties from the quasi-classical partition function

$$Z_f = \int \cdots \int{}' e^{-\beta E(p,r)} \frac{d^2p_1\, d^2r_1 \cdots d^2p_{N_f}\, d^2r_{N_f}}{h^{2N_f}}, \qquad (5.2.3)$$

where the prime refers to "proper Boltzmann counting." Substituting (5.2.1) into (5.2.3), we obtain

$$Z_f = \frac{1}{N_f!}\left[\int \frac{d^2r}{h^2} \int d^2p\; e^{-\beta\epsilon(p,r)}\right]^{N_f} = \frac{1}{N_f!}\left[\frac{A\, \exp\,(\beta\epsilon_0)}{\lambda^2}\right]^{N_f}. \qquad (5.2.4)$$

Using Stirling's approximation, we obtain

$$F_f = -N_f[\epsilon_0 + kT \ln\,(Ae/N_f\lambda^2)]. \qquad (5.2.5)$$

The principal thermodynamic properties are obtained immediately:

$$S_f = -(\partial F_f/\partial T)_{N_f,A} = N_f k[2 - \ln\,(n\lambda^2)], \qquad (5.2.6)$$

$$C_f = T(\partial S_f/\partial T)_{N_f,A} = N_f k, \qquad (5.2.7)$$

$$\phi = -(\partial F_f/\partial A)_{N_f,T} = nkT, \qquad (5.2.8)$$

$$\mu_f = (\partial F_f/\partial N_f)_{T,A} = -\epsilon_0 + kT \ln\,(n\lambda^2). \qquad (5.2.9)$$

The vapor pressure and heat of adsorption are obtained by the same procedure as in the preceding section, and the results here are

$$P = (n/\beta\lambda)\, \exp\,(-\beta\epsilon_0), \qquad (5.2.10)$$

$$q_{st} = \epsilon_0 + \tfrac{3}{2}kT. \qquad (5.2.11)$$

Comparisons between the heat capacities and heats of adsorption of this model and of the Langmuir monolayer were made earlier. Concerning the coverage dependence, we see that in this case the region of linear dependence (known as "Henry's law") extends to arbitrarily high coverage. This is a direct consequence of the lack of any interactions between adatoms in this model and not due to their mobility.

5.2.2 Ideal Quantum Gases†

Effects of quantum statistical interactions become important as temperature is lowered and/or density is increased. Thermodynamic properties can be calculated by following the familiar procedures used in the theory of ideal gases, with allowance for the change in dimensionality, as follows.

The equilibrium thermal occupation numbers of single particle states having energy ϵ are

$$\langle n_{\pm}(\epsilon) \rangle = [e^{\beta(\epsilon-\mu)} \pm 1]^{-1}, \tag{5.2.12}$$

where the upper sign refers to fermions and the lower to bosons. The total number $N_{\pm}$ of particles in any system is the sum of single particle state occupations. If the density of states is a quasi-continuous function of energy $g(\epsilon)$, then $N_{\pm}$ can be written as an integral

$$N_{\pm} = \int \langle n_{\pm}(\epsilon) \rangle g(\epsilon) \, d\epsilon. \tag{5.2.13}$$

Equations (5.2.12) and (5.2.13) are, of course, quite general for all noninteracting systems regardless of dimensionality. Therefore, whatever might be the special character of any system must be due to the form of $g(\epsilon)$. The density of states is directly obtainable from the energy–momentum dispersion relation

$$\epsilon(p) = -\epsilon_0 + p^2/2m \tag{5.2.14}$$

together with the density of momentum states $g(p)$ of 2D phase space

$$g(p) \, dp = \int_A d^2p \, d^2r/h^2 = (2\pi A/h^2)p \, dp, \tag{5.2.15}$$

† Band (3), May (6), McKelvey and Pulver (7), Dash (8).

where A is the total area of the system. The general connection between $g(\epsilon)$ and $g(p)$ is, for any dispersion relation $\epsilon(p)$,

$$g(\epsilon) = g(p) \, d\epsilon/dp. \tag{5.2.16}$$

With (5.2.14), (5.2.15), and (5.2.16), we obtain

$$g(\epsilon) = 2\pi m A/h^2, \qquad -\epsilon_0 < \epsilon < \infty. \tag{5.2.17}$$

Equation (5.2.17) refers to particles without any internal structure, so that each state is completely determined by its kinetic energy ϵ. If there are some internal degrees of freedom then $g(\epsilon)$ would be multiplied by the corresponding degeneracy factors. Thus, if the particles have a nuclear spin s_N there are $g_N = 2s_N + 1$ different states with the same energy, and $g(\epsilon)$ must be multiplied by this factor. In most of the following we will assume that the particles are spinless.

The 2D density of states (5.2.17) differs substantively from the usual 3D relation in that it does not depend on ϵ. This qualitative change is at the heart of several distinctive features of 2D systems. For example it enables one to evaluate the chemical potential in closed form and thereby obtain an analytic expression for the vapor pressure which is valid for all temperatures. Substituting (5.2.12) and (5.2.17) in (5.2.13), we have

$$N_\pm = \frac{2\pi m A}{h^2} \int_{-\epsilon_0}^{\infty} [e^{\beta(\epsilon - \mu)} \pm 1]^{-1} \, d\epsilon = \frac{2\pi m A}{\beta h^2} \int_{-\epsilon_0}^{\infty} \frac{e^{\beta(\mu - \epsilon)} d(\beta \epsilon)}{[1 \pm e^{\beta(\mu - \epsilon)}]}$$
$$= \pm A\lambda^{-2} \ln [1 \pm \exp (\beta(\mu + \epsilon_0))]. \tag{5.2.18}$$

Equation (5.2.18) is analyzed for μ and the two $(\pm)$ solutions are equated to Eq. (5.1.4) for μ_v. Finally, one obtains the vapor pressure relations

$$P_\pm = (\pm)[\exp (\pm n\lambda^2) - 1][\exp (-\beta \epsilon_0)/\beta \lambda^3]. \tag{5.2.19}$$

Equation (5.2.19) is exact at all temperatures and densities. Expanding the exponential factor in powers of $n\lambda^2$, we note that in the classical limit $n\lambda^2 \ll 1$, the Boltzmann approximation (5.2.10) is indeed obtained. The degeneracy parameter increases as the density increases and/or the temperature falls. If $n\lambda^2$ is not much smaller than unity the vapor pressures of fermions and bosons differ in magnitude and

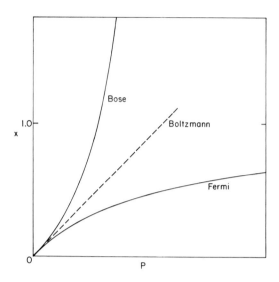

FIG. 5.2   Vapor pressure isotherms of ideal mobile monolayers.

shape. Their density dependences are illustrated in Fig. 5.2. The trends at large n are particularly interesting. For fermions the pressure diverges at a finite density. This is readily understood as a direct result of the Pauli exclusion principle, which prevents multiple occupation of individual momentum states. More specifically, the increasing density of the film causes $n\lambda^2$ eventually to rise above $\beta\epsilon_0$, and the pressure increases exponentially with n. For bosons, on the other hand, the statistical interactions are attractive. As the density is increased the attractions become more important, reducing the pressure below the Boltzmann law. The pressure tends asymptotically at large n to

$$P_+ \text{ (max)} = (\beta\lambda^3)^{-1} \exp\left(-\beta\epsilon_0\right), \tag{5.2.20}$$

and therefore a monolayer of ideal Bose atoms has an infinite capacity at this limiting value of P.

The evaluation of monolayer energy cannot be given in closed form for all temperatures and densities, but it is possible to obtain a universal equation of state relating the 2D pressure ϕ and the kinetic energy of an ideal monolayer correct for all temperatures and both types of statistics:

$$\phi A = E_{\text{kin}} \qquad (5.2.21)$$

(note that the binding energy is excluded). This relation is the analogue of the similarly universal equation for a 3D ideal gas, i.e.,

$$PV = \tfrac{3}{2} E. \qquad (5.2.22)$$

The proof of (5.2.21) follows lines similar to the derivation of (5.2.22), as follows. We begin with the general expression for the energy

$$E_\pm = \int \langle n_\pm(\epsilon) \rangle \epsilon g(\epsilon) \, d\epsilon. \qquad (5.2.23)$$

After substituting (5.2.12) and (5.2.17) for $n_\pm$ and $g(\epsilon)$ and integrating from 0 to ∞ in order to select out only the kinetic energy,

$$E_\pm = \frac{2\pi m A}{h^2} \int_0^\infty \frac{\epsilon \, d\epsilon}{1 \pm e^{\beta(\epsilon - \mu)}} \qquad (5.2.24)$$

Integrating by parts, we obtain

$$E_\pm = A\lambda^{-2} \left\{ (\mp) \epsilon \ln \left[1 \pm e^{\beta(\mu - \epsilon)} \right] \Big|_0^\infty \pm \int_0^\infty d\epsilon \ln \left[1 \pm e^{\beta(\mu - \epsilon)} \right] \right\}$$
$$= \pm A\lambda^{-2} \int_0^\infty d\epsilon \ln \left[1 \pm e^{\beta(\mu - \epsilon)} \right]. \qquad (5.2.25)$$

Equation (5.2.25) is now compared with the thermodynamic potential Ω. For noninteracting particles in any distribution of discrete single particle levels, the potentials for fermions and bosons are

$$\Omega_\pm = \mp kT \sum_i \ln \left\{ 1 \pm \exp \left[\beta(\mu - \epsilon_i) \right] \right\}. \qquad (5.2.26)$$

In a quasi-continuous distribution of states, the sum can be replaced by an integral weighted by $g(\epsilon)$. With the 2D distribution (5.2.17) and ignoring the binding energy, (5.2.26) becomes

$$\Omega_\pm = \mp A\lambda^{-2} \int d\epsilon \ln \left[1 \pm e^{\beta(\mu - \epsilon)} \right]. \qquad (5.2.27)$$

Equation (5.2.27) is just the negative of (5.2.25); since the limiting expression for 2D matter is $\Omega = -\phi A$ [see Eq. (5.2.17) of Chapter 4], the equation of state (5.2.21) is proved.

Although there is no simple analytic solution for the energy at arbitrary temperatures and densities, May (6) derived an exact relation between the energies of Bose and Fermi systems under conditions of identical temperature, density, particle mass, and spin. May began with the expression for energy as we have it in (5.2.24), expanded the denominator as a power series in $\beta\epsilon$ and integrated term by term. Comparing expressions for $+$ and $-$, we have the result

$$E_{\mathrm{FD}} = E_{\mathrm{BE}} + \tfrac{1}{2}NkT_0, \qquad (5.2.28)$$

where T_0 is the temperature at which $n\lambda^2 = 1$. Therefore, since the energies differ by only a constant,

$$C_{\mathrm{FD}}(N, T) = C_{\mathrm{BE}}(N, T) \qquad (5.2.29)$$

for all temperatures and densities. This is an extremely surprising result in view of the striking difference between P_{FD} and P_{BE}. In 3D gases there are qualitative differences between Fermi and Bose gas heat capacities. The general relation (5.2.29) shows that there cannot be any distinctive feature such as a Bose–Einstein condensation peak or cusp in noninteracting 2D systems, in contrast to the 3D situation. In fact there is no condensation at any finite T (9). Condensation implies a macroscopic occupation of the lowest single particle level of the system, i.e.,

$$\langle n_{\mathrm{BE}}(\epsilon = \epsilon_0) \rangle = O(N).$$

This will occur when μ reaches the value ϵ_0; solving (5.2.18) for μ, one finds that it requires either $T = 0$ or $n = \infty$. The overall temperature dependence of $C_{\pm}$ is rather like the familiar shape of an ideal Fermi gas, as is illustrated in Fig. 5.3. [The calculation of C_{FD} and other thermodynamic functions of ideal Fermi gases can be made for arbitrary N, T by means of numerical tables of the Fermi integral (10, 11)]. Both curves are smooth, concave toward the temperature axis, and linear in T near $T = 0$. The extreme low temperature behavior of C_{FD} can be derived by the same techniques as is conventional for the degenerate Fermi gas. Using May's equality the same expressions hold for C_{BE}. The derivation is quite straightforward: here we merely state the result:

$$C = (\pi^2/3)Nk(T/T_0); \qquad T \ll T_0, \qquad (5.2.30)$$

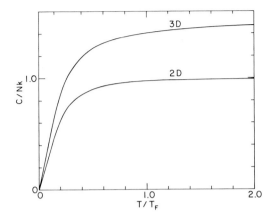

FIG. 5.3   Specific heats of ideal 2D and 3D spin $\frac{1}{2}$ Fermi gases.

where

$$kT_0 = (h^2/2\pi m g_N)(N/A).$$

T_0 is the same characteristic temperature as is given in (5.2.28) provided that when the nuclear spin s is nonzero, we modify the definition of the wavelength to contain the spin statistical factor $g_N = 2s + 1$ according to $\lambda \equiv h(2\pi m k T g_N)^{-1/2}$. The corresponding formulas for the 3D Fermi gas at very low temperatures $T \ll T_F$ are

$$C(\text{3D Fermi gas}) = (\pi^2/2)Nk(T/T_F);$$

where

$$kT_F = (h^2/4\pi^2 m g_N)(6\pi^2 N/g_N V)^{2/3}. \tag{5.2.31}$$

Thus the characteristic degeneracy temperature T_0 of ideal monolayers of *either statistics* plays a role similar to that of the degeneracy temperature T_F in a 3D Fermi gas.

There is an interesting difference between the 2D and 3D low temperature heat capacities. In 3D the heat capacity depends on both N and $(N/V)^{2/3}$, but in 2D, $C(T \to 0)$ depends only on A and not on N. An additional noteworthy distinction is that in 2D the linear region is restricted to much lower relative temperatures, as one can see from Fig. 5.3. At the value $T = T_0$, the monolayer heat capacity is only a few percent lower than the classical high temperature limit Nk.

5.3 THERMODYNAMICS OF THE BAND MODEL

The basis of the surface band model has been outlined in Chapter 2. We shall now study the statistical properties of bosons and fermions in a succession of simple band structures, showing the influence of simple parameters such as bandwidths and energy gaps.

5.3.1 An Isolated Low-Lying Band†

We first assume that the energy spectrum in the range of interest consists of only a single isolated band. This case, as we shall see later, approximates a situation in which there is a relatively narrow lowest band, well separated from higher bands, for values of kT much smaller than the energy gap. The density of states is some $g(\epsilon)$; for simplicity we will assume the standard band approximation, so that $g(\epsilon)$ in the band is independent of energy and direction along the surface. Ignoring the question of binding to the surface in this first example, we have from Chapter 2, Section 2.3,

$$g(\epsilon) = 2\pi m^* A/h^2 = N_s/\delta, \qquad 0 \lesssim \epsilon \lesssim \delta,$$
$$= 0, \qquad\qquad\qquad \epsilon > \delta. \qquad (5.3.1)$$

Now let us calculate the free energy in the Boltzmann approximation, which is valid as long as the occupation number $\langle n(\epsilon) \rangle \ll 1$ for each energy level (the conditions for which the inequality is obeyed are established below). Using (5.3.1) one can obtain the partition function by straightforward integration. The free energy and chemical potential are

$$F = -NkT \ln \left[(eN_s/N\beta\delta)(1 - e^{-\beta\delta}) \right], \qquad (5.3.2)$$

$$\mu = -kT \ln \left[(N_s/N\beta\delta)(1 - e^{-\beta\delta}) \right]. \qquad (5.3.3)$$

From (5.3.3) we obtain the criterion for the region of validity of the Boltzmann approximation. Since $\langle n(\epsilon) \rangle = [e^{\beta(\epsilon-\mu)} \pm 1]^{-1} \ll 1$ when $e^{-\beta\mu} \gg 1$, we see that the Boltzmann condition requires

$$(N_s/N)[(1 - e^{-\beta\delta}/\beta\delta] \gg 1.$$

† From Dash and Bretz (5) and Dash (8).

This condition is satisfied for low coverages *and* "high" temperatures

$$N \ll N_s, \qquad \beta\delta \ll 1.$$

The Boltzmann condition can also be stated in terms of a degeneracy parameter $n(\lambda^*)^2$, where λ^* is the thermal de Broglie wavelength for particles of effective mass m^*; then one obtains an inequality resembling the ordinary 2D criterion, i.e.,

$$n(\lambda^*)^2 \ll 1.$$

The temperature dependence of S and of the heat capacity are particularly interesting. At "very high temperatures," $\beta\delta \ll 1$, expanding the exponential in (5.3.2) to first order in $\beta\delta$, we find that the entropy $S = -(\partial F/\partial T)_{N,N_s}$ has a constant value

$$S(kT \gg \delta) = Nk \ln (eN_s/N), \qquad (5.3.4)$$

and hence the heat capacity vanishes. Equation (5.3.4) is the expression for a random distribution of N noninteracting Boltzmann particles among N_s equivalent states in the thermodynamic limit. There is no way of telling from (5.3.5) anything concerning the dynamical nature of the states in question; if instead of the band model we had begun instead with a localized model and allowed multiple occupation of sites, the same result would be obtained.

At somewhat lower temperatures the occupation of higher levels begins to fall, decreasing S and giving a nonzero heat capacity. Expanding $e^{-\beta\delta}$ to third order in $\beta\delta$, one can show that the heat capacity $C = T(\partial S/\partial T)_{N,N_s}$ begins to rise as

$$C = \tfrac{1}{12}Nk(\delta/kT)^2. \qquad (5.3.5)$$

Equation (5.3.5) is identical to the high temperature tail of a "Schottky peak." It is characteristic of any system having an isolated group of low-lying energy levels, at relatively high temperatures.

Now suppose the temperature is quite low such that $e^{-\beta\delta} \ll 1$, and that the fractional coverage is sufficiently low that the Boltzmann approximation is still valid. Here we find that $C = Nk$. This 2D classical gas result is not surprising when we consider that the system can indeed be represented as a 2D gas of Boltzmann particles of mass m^*, if the alternative expression for the density of states is used (Eq.

(5.3.1)). The "low temperature" condition $e^{-\beta\delta} \ll 1$ on which the result depends guarantees that the distribution is not markedly affected by the fact that $g(\epsilon)$ falls to zero at $\epsilon > \delta$.

At still lower temperatures the Boltzmann approximation will eventually become invalid, and then one must take account of statistical interactions. These will follow the same type of behavior as the ideal 2D gas, except that m^* replaces m. Thus, the heat capacity decreases and eventually goes to 0 linearly with T.

The preceding analysis is based on the condition $N/N_s \ll 1$, allowing the use of Boltzmann statistics down to temperatures much lower than δ/k. If N/N_s is not very small then the Boltzmann approximation cannot be used at any temperature. One consequence of finite coverage is that the heat capacity never reaches the quasiclassical limit Nk, but has a rounded maximum at a lower value. The detailed shape of the anomaly depends upon the fractional coverage as well as statistics and

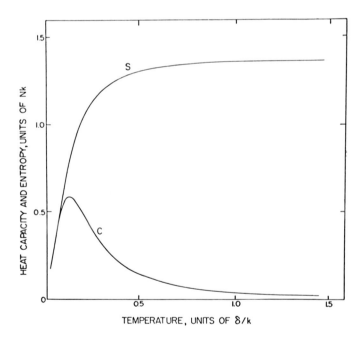

FIG. 5.4   Specific heat and entropy due to a half-filled solitary tunneling band; spinless fermions with uniform effective mass over the band width δ.

is not obtainable analytically. However, we can recognize that certain gross features are preserved: C vanishes in the limit $\beta\delta = 0$, rises as T^{-1} in the region $\beta\delta \ll 1$, has a maximum value at $\beta\delta \approx 1$, and decreases linearly to 0 at $T = 0$. Numerical results for $N/N_s = \frac{1}{2}$ are illustrated in Fig. 5.4.

5.3.2 Spectrum with an Energy Gap†

The next example is a system of two bands, as shown in Fig. 5.5. Such a band structure represents an isolated "tunneling band" lying below a continuum of free translational states. First we will examine the case when the energy gap Δ is much greater than the tunneling band width δ. In this situation the thermal effects of the band and the continuum occur at quite different temperatures. At very high temperatures, $kT \gg \Delta$, the influence of the lower band is negligible since almost all of the particles are distributed among the higher regions of the spectrum. If the density of states in the continuum has the characteristic form for 2D free particle states, i.e., $g(\epsilon)$ independent of ϵ, then the thermal properties are those of a 2D gas on a smooth surface.

† From Dash and Bretz (5) and Dash (8).

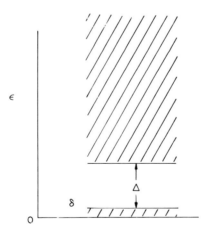

FIG. 5.5 Simple two-band spectrum consisting of a low-lying tunneling band separated from a higher thermal band.

Thus, if in addtion to the condition $kT \gg \Delta$ the degeneracy parameter is such that $n(\lambda_c{}^*)^2 \ll 1$ (where $\lambda_c{}^*$ is computed for the effective mass $m_c{}^*$ in the continuum), then the film is "classical" and $C = Nk$. As the temperature is decreased the heat capacity remains at this value until either statistical interactions or effects due to the gap and lower band are felt. Details of the behavior now depend on the density of states in the continuum and the gap Δ as well as N/N_s, δ, and statistics. It is, nevertheless, still possible to describe certain general features. If N/N_s is relatively low the influence of the gap is felt before (i.e., at higher temperatures) that of statistics. As kT falls toward the value Δ the absence of states in the gap causes the entropy to decrease more rapidly than in the strictly 2D case, and hence C rises above Nk and has a rounded peak near where $S(T)$ changes most rapidly. The entropy continues to fall with decreasing T, toward its asymptotic value (5.3.4) for an isolated band at "high temperatures" $kT \gg \delta$. In the region $\delta \ll kT \ll \Delta$ the heat capacity contributions of both bands are very small. Then, as T is lowered still further, the effect of the lower band is seen. A particularly interesting feature is the virtual absence of any

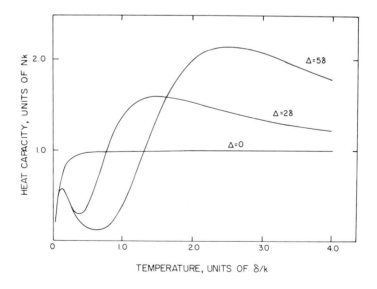

FIG. 5.6 Heat capacity of spinless fermions at coverage $x = \frac{1}{2}$ in the two-band spectrum of Fig. 5.5.

heat capacity in the intermediate region, with no suggestion of any anomaly lying at very low temperatures. Yet the existence of the low temperature anomaly is demanded by the third law (in its "strong form" as, for example, in Landau and Lifshitz, "Statistical Physics"— ". . . the entropy of any body vanishes at absolute zero."). The low temperature peak occurs as the noninteracting model eventually removes the configuration entropy. Of course, if adatoms interact there are other ways of removing the degeneracy, but here we assume that there are no interactions.

If the energy gap is not much greater than the lowest band width, the high and low temperature regimes are less distinct as illustrated in Fig. 5.6. The standard 2D result, with no peaks at all, is obtained as $\Delta \to 0$.

5.3.3 Other Band Structures†

Whenever it has been necessary to adopt a particular form for $g(\epsilon)$, we have assumed up to this point the standard band approximation, i.e., that $g(\epsilon)$ is uniform in each band. The approximation is quite convenient as an illustration but realistic structures are more complex (12, 13). In Fig. 5.7, for example, is shown $g(\epsilon)$ computed for the low-lying states of He adsorbed on the basal plane surface of graphite plated with a monolayer of Xe (16). From the irregular nature of $g(\epsilon)$ one might expect that the temperature dependence of $C(T)$ would depart markedly from a smooth 2D result, but Fig. 5.8 shows that it does not (16). This is because the spectrum is effectively averaged over an energy width kT; hence any fine-grained structure in $g(\epsilon)$ can only show up at very low temperatures. In principle one can make use of the increased resolution at small kT by carrying out a series of heat capacity measurements on adsorbed fermions at very low temperatures (12). When $\beta\mu \gg 1$ the specific heat of any system of fermions can be expanded in the usual asymptotic series:

$$C = \frac{\pi^2 k^2 T}{3} \left[\frac{\partial(\epsilon g(\epsilon))}{\partial \epsilon} \right]_{\epsilon=\mu} + \cdot \cdot \cdot . \tag{5.3.6}$$

† From Dash and Bretz (5), Dash (8), Bretz (12), Widom (13), Milford and Novaco (14), Hagen *et al.* (15), Novaco and Milford (16).

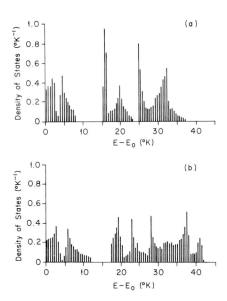

FIG. 5.7   Density of states for (a) ^{3}He and (b) ^{4}He adsorbed on Xe-plated graphite, calculated by Novaco and Milford (16). E_0 is the ground state energy in each case.

Thus the slope of the linear region near $T = 0$ depends on the density of states at the top of the Fermi distribution. A complete mapping of $g(\epsilon)$ could be made by a systematic study of $C(T \rightarrow 0)$ as a function of coverage. Analogous studies have been made in metallic alloys to determine the density of states of conduction electrons.

There is a similar low temperature expansion that can be made for the vapor pressure of a Fermi monolayer (8). In this case it is $g(\epsilon)$ itself that is measured rather than the mixed term in Eq. (5.3.6). The pressure dependence is obtained from applying the standard low temperature series expansion of the Fermi integral to the expression for particle number N:

$$N = \int_0^\mu \frac{g(\epsilon)\, d\epsilon}{e^{\beta(\epsilon - \mu)} + 1} = \int_0^\mu g(\epsilon)\, d\epsilon + \frac{\pi^2}{6} (kT)^2 \left[\frac{\partial g(\epsilon)}{\partial \epsilon} \right]_\mu + \cdots .$$

We now differentiate with respect to particle number at constant T,

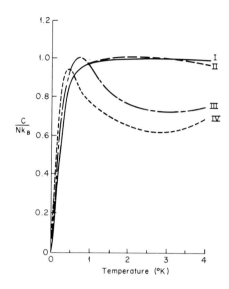

FIG. 5.8 Specific heat of ³He adsorbed on (I) a plane surface, (II) the basal plane of graphite, (III) a close-packed Ar layer on a smooth surface having the average binding energy of Cu, and (IV) Xe-plated graphite; calculated by Novaco and Milford (16). Adatom interactions are neglected, but statistical effects are included. Film density is 0.025 atoms (Å)⁻², or about ¼ monolayer.

which yields, to first order,

$$g(\mu) = (\partial N/\partial \mu)_T. \tag{5.3.7}$$

The coverage dependence of μ can be expressed in terms of changes in the vapor pressure P. Using (5.1.4), we obtain

$$g(\mu) = [kT(\partial \ln P/\partial N)_T]^{-1}. \tag{5.3.8}$$

Hence, the effective density of states of noninteracting fermions could be mapped directly by a vapor pressure isotherm at low temperatures.

5.4 CRITERIA FOR MOBILITY

The band theory provides a simple scheme for studying effects of mobility on thermal properties, just by varying one or two parameters such as band width and energy gap. It allows us to follow the thermal changes caused by a gradual evolution from a "completely localized" (zero bandwidth) model to a "completely mobile" (zero energy gap) model. Since the mobility in any real system is intermediate to these two extremes, we need to develop quantitative gauges of mobility, and if these criteria are to be useful they must be given in terms of experimental quantities. This suggests that there might be several different mobility "scales," one for each type of measurement. These scales need not be proportional or simply related to each other since the different thermodynamic properties of a system depend individualistically on the momenta of the single particles. It is even possible that a system which behaves as a "localized" monolayer with respect to one type of measurement can appear quite mobile in other experiments under the same thermodynamic conditions.

Particularly interesting are the criteria for mobility associated with heat capacity and vapor pressure. In the simple two-band model, the heat capacity has a maximum at $kT \approx \Delta$ and then a decrease proportional to $e^{-\Delta/kT}$ at lower temperatures. This behavior is similar in form to that of Hill's approximation (4) in the region of the "mobile-localized transition," and hence according to this calculation it would seem that the atoms become more completely "localized" at lower and lower temperatures. But more correctly in terms of the band model, we understand that the heat capacity maximum is simply due to the rapid depopulation of the continuum states. When a particle drops down into the tunneling band it undergoes a decrease in mobility (in the sense of a decrease in kinetic energy) but as long as the band has a finite width the particle mobility never decreases to zero. The dwell time of an adatom on any particular site is related to the bandwidth by $\tau_s \approx \hbar/\delta$, and this condition holds down to $T = 0$. Nevertheless, the lowest band is not evidenced in the heat capacity until the temperature is lowered to the neighborhood of some temperature T_t, where the low temperature peak appears. From the approximate location of the low temperature peak in relation to the bandwidth we obtain a criterion for temperatures at which the tunneling should be evident in the heat capacity (5):

$$T \gtrsim T_t \approx \hbar/k\tau_s. \qquad (5.4.1)$$

Thus all (noninteracting) monolayers have characteristically "mobile" heat capacities at sufficiently low temperatures. To distinguish this region from conventional thermal mobility at high temperatures, it can be called "quantum mobility." If the two regions are well separated, then a monolayer will appear to be "localized" at temperatures lying between δ/k and Δ/k, and to have two "localized-mobile transitions."

All of this discussion has been predicated on the assumption that the film is always in equilibrium, which requires that the heat capacity measurement be made slowly enough for the system to be fully thermally relaxed at all stages. The internal equilibrium depends on adatom–adatom interactions and energy exchanges, and these must be superimposed on the simplified noninteracting model of the present discussion. At best we must require, if the band model is to retain any validity, that the interactions are sufficiently weak so that the translational states do not decay too rapidly (in less than one or two wavelengths). This condition requires that the average energy of interaction be weaker than the bandwidth. When the condition is translated into times, it becomes $\tau_{rel} > \tau_s$. Then collecting the inequalities, we have the criterion for detection of quantum mobility in a specific heat measurement,

$$\tau_{exp} \gg \tau_{rel} > \tau_s ,\qquad (5.4.2)$$

simultaneously with (5.4.1).

Now let us consider the influence of mobility on the vapor pressure. In this case there is a natural interval which establishes a time scale for the film; the lifetime τ_{des} of an atom before desorption into the vapor phase. Intuitively, we can argue that if τ_{des} is shorter than τ_s for escape from the site by translation along the surface, the film must appear "localized" in vapor pressure measurements. This is indeed the case, as one can show by more careful arguments (8). However, the distinction is fairly academic, since there are no major differences between the isotherms of mobile and localized monolayers. Here again we emphasize that the qualitative differences between the isotherms of the Langmuir and 2D gas models are not due to differences in mobility but are consequences of the site exclusion (i.e., interactions) in the Langmuir model *vis-à-vis* the ideality of the gas model. In contrast to the heat capacity, the vapor pressure is a quite insensitive gauge of monolayer mobility.

There are a number of transport coefficients which should provide more direct measurements of mobility. Since they are strongly governed by particle interactions we postpone their discussion to subsequent chapters that deal with condensed phases.

5.5 RELEVANT EXPERIMENTS

Although the thermodynamics of mobile and localized monolayers has been a topic of interest for over half a century, there have been very few experimental tests of the theory. This is because of a combination of constraints and technical difficulties. To approximate the noninteracting models the density of adsorbed atoms must be quite low, and this causes problems of detectability, whatever the experimental quantity and technique. Also, the differences between localized and mobile adsorption involve relatively small differences in energy (compared to the heat of adsorption, for example), which places particularly stringent demands on the uniformity of the substrate. For a really thorough-going test, one would want to compare theory and experiment for a system having a fairly structured temperature-dependent property, such as the well-modulated heat capacities illustrated in Fig. 5.6. Such a comparison has not yet been done, but a partial test of a simpler system has been successful. The test system is He adsorbed on basal planes of graphite, which as noted in the preceding section, was predicted by Hagen et al. (15, 16) to have low temperature heat capacities which are virtually indistinguishable from He on a smooth plane. Experimental results for low coverage ^{3}He and ^{4}He adsorbed on graphite (17–19) are in close agreement with the predictions, at least over the temperature range 2–4°K above the region of strong gas imperfection (see Figs. 6.2 and 6.3). This agreement is highly satisfying, yet it must be qualified on two counts. First, the agreement is in a sense one-sided, in that the data show that the bandwidth is not much narrower than the theoretical value, for then the heat capacity would show a dip due to the energy gap. But if the substrate is actually much smoother than the calculation there would still be good agree-

ment with the strictly 2D results. A second and more important quali-
fication is that the effects of interactions are actually quite important
at the experimental coverages. The combined effects of the adatom-
adatom and adatom–substrate interactions are sufficiently strong, even
at these relatively low densities, so that the mobility which is evident
in the heat capacity is probably not strictly a single particle property,
but is instead a collective property of the entire system. These points
are discussed *in extenso* in Chapter 8.

REFERENCES

1. I. Langmuir, *J. Amer. Chem. Soc.* **40**, 1361 (1918).

2. R. H. Fowler and E. A. Guggenheim, "Statistical Thermodynamics," Chapter X. Cambridge Univ. Press, London and New York, 1939.

3. W. Band, "An Introduction to Quantum Statistics," pp. 165–173. Van Nostrand-Reinhold, Princeton, New Jersey, 1955.

4. T. L. Hill, *J. Chem. Phys.* **14**, 441 (1946).

5. J. G. Dash and M. Bretz, *Phys. Rev.* **174**, 247 (1968).

6. R. M. May *Phys. Rev.* **135**, A1515 (1964).

7. J. P. McKelvey and E. F. Pulver, *Amer. J. Phys.* **32**, 749 (1964).

8. J. G. Dash, *Phys. Rev.* **A1**, 7 (1970).

9. M. F. M. Osborne, *Phys. Rev.* **76**, 396 (1949).

10. P. Rhodes, *Proc. Roy. Soc. (London)* **A204**, 396 (1950).

11. A. C. Beer, M. N. Chase, and P. F. Choquard, *Helv. Phys. Acta* **28**, 529 (1955).

12. M. Bretz, *Phys. Rev.* **184**, 162 (1969).

13. A. Widom, *Phys. Rev.* **185**, 344 (1969).

14. F. J. Milford and A. D. Novaco, *Phys. Rev.* **A4**, 1136 (1971).

15. D. E. Hagen, A. D. Novaco, and F. J. Milford, *in* "Adsorption–Desorption Phenomena" (F. Ricca, ed.). Academic Press, New York, 1972.

16. A. D. Novaco and F. J. Milford, *Phys. Rev.* **A5**, 783 (1972).

17. M. Bretz and J. G. Dash, *Phys. Rev. Lett.* **26**, 963 (1971).

18. D. C. Hickernell, E. O. McLean and O. E. Vilches, *Phys. Rev. Lett.* **28**, 789 (1972).

19. M. Bretz, J. G. Dash, D. C. Hickernell, E. O. McLean, and O. E. Vilches, *Phys. Rev.* **A8**, 1589 (1973).

6. Two-Dimensional Imperfect Gases and Phase Condensation

The noninteracting models of the previous chapter might be suitable approximations to real films if the coverage is low enough, but this situation is rarely encountered in experiments. For practical reasons the film density is usually an appreciable fraction of a completed layer, which is tantamount to a condition that interactions are important. Accordingly, one might expect to encounter a variety of regimes analogous to those of bulk matter when interactions come into play; deviations from ideal gas behavior, becoming stronger with increasing density and/or decreasing temperature, and condensation into liquid and solid phases. Such behavior is indeed observed, although when the substrate structure is comparable in importance to the adatom interactions, the film regimes become strongly modified and novel characteristics appear. These complications are discussed in Chapter 8. In the present chapter we neglect the surface texture for the most part.

By way of introduction, we note that striking evidence of phase changes in monolayer films in semisoluble oil films floating on water was obtained as long ago as the 19th century (1). In these studies the spreading pressure ϕ of the film is directly measured as a force exerted on a sliding barrier and is observed to change as the barrier pushes the film into a smaller area of the surface. With a variety of oil films the experimental ϕ–A diagrams are remarkably similar to the familiar P–V diagrams of bulk matter (2, 3). For films on solid surfaces there are several thermodynamic quantities that are more suitable than ϕ for direct measurement, and all of them have in various instances given strong indications of phase changes. The vapor pressure, for example, has been employed for particularly thorough and detailed explorations of the phase changes in light molecular gas monolayers adsorbed on exfoliated graphite (4). A set of such vapor pressure isotherms is shown in Fig. 6.1. These results and other experimental measurements dealing with adatom–adatom interactions and phase condensation are discussed in Section 6.3 after an outline of the simple theory. In this first approach to the problem of interacting monolayers we ignore the substrate structure and treat the monolayer as a classical 2D gas. The approximation allows us to follow along the lines of the familiar theory of imperfect gases merely by reducing the dimensionality and adding a term for the substrate binding energy. The details of calculation and the results are discussed at some length before turning to a discussion of the full quantum mechanical virial expansion and the detailed calcu-

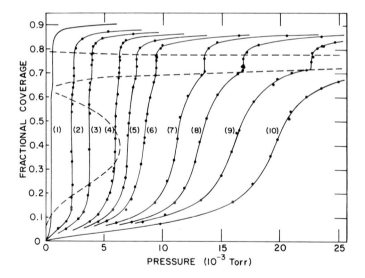

FIG. 6.1 Vapor pressure isotherms of Kr on exfoliated graphite during the formation of the first layer, as reported by Thomy and Duval (4). Temperatures of the isotherms (°K) are: (1) 77.3; (2) 82.4; (3) 84.1; (4) 85.7; (5) 86.5; (6) 87.1; (7) 88.3; (8) 89.0; (9) 90.1; (10) 90.9.

lations of the second virial coefficients of ^{3}He, ^{4}He, Ar, and Ne. The calculations are compared with experimental heat capacities for each type of molecule at submonolayer coverage on graphite substrates, with striking agreement in the case of the He isotopes.

6.1   WEAK INTERACTIONS; BOLTZMANN APPROXIMATION†

The substrate is assumed to be a smooth attracting plane, with single particle binding energy ϵ_0. With adatom–adatom interactions the energy of a monolayer of N adatoms is

$$E(\mathbf{p}, \mathbf{r}) = -N\epsilon_0 + \sum_i^N (p_i{}^2/2m) + U(\mathbf{r}_1, \mathbf{r}_2, \ldots, \mathbf{r}_N). \quad (6.1.1)$$

At relatively high temperatures the system can be treated in the Boltzmann approximation, which permits a complete factoring of the partition function into kinetic and configurational terms, corresponding to the particular form of the terms in $E(\mathbf{p}, \mathbf{r})$. Thus, if we define Z_0 to be the partition function of the ideal noninteracting monolayer, the total partition function Z_f in the presence of interactions is

$$Z_f = Z_0 Q(\beta, N, A), \quad (6.1.2)$$

$$Q = A^{-N} \int \cdots \int e^{-\beta U} d^2r_1 \cdots d^2r_N \quad (6.1.3)$$

is the "configurational factor" due to the interactions. If U is a sum of pair interactions of the form $u_{ij}(|\mathbf{r}_i - \mathbf{r}_j|)$, then the integral in (6.1.3) can be transformed, in the low density approximation, to a sum over pair collisions; these details are identical to the classical theory of imperfect gases (11) except for the reduced dimensionality. The resulting approximation for the Helmholtz free energy is

$$F_f \cong F_0 - kT \ln \{1 - [N(N-1)/A]B_f(T)\}$$

$$\cong F_0 + (N^2 kT/A)B_f(T), \quad (6.1.4)$$

† Volmer (5), Fowler and Guggenheim (6), Cassel (7), Hill (8), Steele and Halsey (9), Steele (10).

where $B_f(T)$ is the second virial coefficient due to pairs of particles;

$$B_f(T) \equiv \tfrac{1}{2} \int_0^\infty [1 - \exp(-\beta u_{12})] \, d^2 r_{12}. \qquad (6.1.5)$$

$B_f(T)$ has an analytic form similar to the usual 3D coefficient, differing only in the dimensionality of the volume element.

Before proceeding to an evaluation of the second virial coefficient explicitly in terms of the pair interaction it is useful to list the effects of a nonspecific $B(T)$ on the various thermodynamic quantities. Using standard relations derived in Chapter 4, we obtain the following general formulas from Eq. (6.1.4):

$$S_f/N_f k = 2 - \ln(n\lambda^2) + n\beta^2 (d/d\beta)(B/\beta), \qquad (6.1.6)$$

$$C_f/N_f k = 1 - n\beta^2 \, d^2 B/d\beta^2, \qquad (6.1.7)$$

$$\mu_f = -\epsilon_0 + (1/\beta) \ln(n\lambda^2) + 2nB/\beta, \qquad (6.1.8)$$

$$\phi = (n/\beta)(1 + nB), \qquad (6.1.9)$$

$$P = (nkT/\lambda) \, e^{-\beta \epsilon_0} \, e^{2nB}, \qquad (6.1.10)$$

$$q_{st} = \epsilon_0 + (3/2\beta) - 2n \, dB/d\beta. \qquad (6.1.11)$$

To approximate B for typical systems we recognize that pair interactions are strongly repulsive at close approach and weakly attractive at larger distances. If we model this behavior as an abrupt crossover from strong repulsion to weak attraction at some characteristic distance σ, then $B_f(T)$ can be written as a sum of two terms, which can be estimated semiquantitatively:

$$B_f(T) = \tfrac{1}{2} \int_0^\sigma [1 - \exp(-\beta u_{12})] \, d^2 r_{12} + \tfrac{1}{2} \int_0^\infty [1 - \exp(-\beta u_{12})] \, d^2 r_{12}.$$
$$(6.1.12)$$

In the close region, u_{12} is large and positive, so that the first term is approximately a constant (the effective "collision area"), i.e.,

$$\tfrac{1}{2} \int_0^\sigma [1 - \exp(-\beta u_{12})] \, d^2 r_{12} \simeq \tfrac{1}{2} \int_0^\sigma d^2 r_{12} \equiv b. \qquad (6.1.13)$$

The attractive term at relatively high temperatures $\beta u_{12} \ll 1$ is approximated by expanding the exponential to first order:

$$\tfrac{1}{2} \int_\sigma^\infty [1 - \exp(-\beta u_{12})] \, d^2 r_{12} \simeq (\beta/2) \int_\sigma^\infty u_{12} \, d^2 r_{12} \equiv -\beta a. \qquad (6.1.14)$$

The parameter a is bounded if the attraction falls off more rapidly than an inverse square law. The sign convention is such that $a > 0$. Thus we obtain the high temperature approximation

$$B_f(T) = b - a/kT. \qquad (6.1.15)$$

The attractive and repulsive terms are sensed in distinctive ways by each of the thermal functions. For example, the entropy is reduced by an amount proportional to the molecular area b, but is completely insensitive to the attractive term. The heat capacity is unaffected by either term. We can understand the decrease in entropy as due to the reduction in "free area," Nb being occupied by the atoms. Since the change in S is independent of T the specific heat is unchanged from its ideal value. This is only true in the classical approximation. In Section 6.4 it is demonstrated that when quantum effects become important the heat capacity is affected by both repulsive and attractive interactions. The vapor pressure equation (6.1.10) is particularly interesting in that it responds to both a and b to varying degrees, depending on temperature. At very high temperature, where $b > a\beta$ and $B > 0$, the vapor pressure rises with coverage as $n \exp (\text{const } n)$, and this behavior, in the region of moderate coverage, resembles the Langmuir and noninteracting fermion isotherms which exhibit saturation at some definite monolayer density. At lower temperatures B changes sign (for 3D gases the temperature at which B changes sign is known as the "Boyle temperature"). At the crossover temperature, $B = 0$ and the monolayer vapor pressure obeys the ideal Boltzmann isotherm. At lower temperatures the vapor pressure falls below the ideal curve. As n increases dP/dn decreases due to the tendency of the adsorbed atoms to attract additional atoms to the surface, and this is reflected by the increase of q_{st} with increasing coverage.

The approximation outlined in these paragraphs amounts to a truncated virial expansion for the film; any subsequent connection between the properties of the film and those of the vapor, so as to derive expressions for P and q_{st}, are made by equating chemical potentials of the interacting monolayer and the (noninteracting) vapor. The procedure, although justified in practical cases, is somewhat inconsistent. The vapor and film are both subject to interactions with the substrate to varying degrees. As noted in Chapter 4, it is possible to treat the combination of "film" and "vapor" as a single heterogeneous imperfect gas, with virial corrections due to interactions among the gas molecules themselves and between them and the surface. This approach was originated by Steele and Halsey (9), and subsequently carried forward in considerable detail in both theory and experiment (10).

6.2 TWO-DIMENSIONAL VAN DER WAALS GASES

The theory in Section 6.1 is valid only for low densities, where multiple collisions are unimportant. Nevertheless, the low density approximation can be converted into a form leading to a "2D van der Waals" equation which, just as in the 3D case, can serve as an interpolation formula into the region of high density. Thus, substituting (6.1.14) into (6.1.9), the 2D pressure equation can be put into the form

$$(\phi + n^2 a)(1 - nb) = nkT. \qquad (6.2.1)$$

Equation (6.2.1) exhibits the same type of loop in the ϕ-n plane as is shown in the P-ρ plane by the ordinary 3D van der Waals formula and which, by the Maxwell construction, is interpreted as gas–liquid condensation. The critical parameters of the 2D equation, corresponding to the simultaneous conditions

$$(\partial \phi / \partial n)_T = 0; \qquad (\partial^2 \phi / \partial n^2)_T = 0 \qquad (6.2.2)$$

are, in terms of the repulsive and attractive components of the virial coefficient, formally identical to the 3D parameters:

$$n_c = b/3, \qquad kT_c = 8a/27b, \qquad \phi_c = a/27b^2. \qquad (6.2.3)$$

Using these critical relations, Eq. (6.2.1) can now be written in dimensionless form in terms of reduced variables

$$\phi_r \equiv \phi / \phi_c, \qquad t \equiv T/T_c, \qquad n_r \equiv n/n_c,$$
$$(\phi_r + 3n_r^2)(3 - n_r) = 8n_r t. \qquad (6.2.4)$$

The 3D equivalent of Eq. (6.2.4), which has an identical appearance, is the basis for the law of corresponding states, and it suggests that there might be a comparably successful law for adsorbed monolayers. Indeed, there are several experimental studies that are consistent with it although it is by no means universal. Disagreements are not at all surprising since the theory ignores substrate structure and substrate-mediated interactions. Any appreciable modulation of the surface potential will affect both the mobility and the spatial distribution of adatoms, each type of gas atom being influenced in more specific ways than can be accounted for by a simple scaling of its pair potential. Further remarks along these lines are made after a review of experiments.

Keeping the above qualifications in mind, it is interesting to pursue the van der Waals theory a bit further, to compare the critical point parameters of monolayer films with those of the bulk phases. Since the relations listed in (6.2.3) are independent of dimensionality, the ratios of 2D and 3D critical coefficients are obtained simply from the expressions for the constants, b and a given in Eqs. (6.1.12) and (6.1.13), together with their equivalent 3D formulas. If the range dependence of the attractive potential is as $r^{-\nu}$ ($\nu > 3$), then it is easy to show that the critical temperatures are in the ratio

$$T_c(2D)/T_c(3D) = \tfrac{2}{3}[(\nu - 3)/(\nu - 2)], \qquad (6.2.5)$$

which equals $\tfrac{1}{2}$ for $\nu = 6$. The ratio of critical densities in dimensionless form is

$$[n_c(2D)]^{3/2}/n_c(3D) = \tfrac{4}{3}(2/\pi)^{1/2} = 1.06, \qquad (6.2.6)$$

independent of ν.

6.3 EXPERIMENTAL RESULTS

The measurements of Thomy and Duval illustrated in Fig. 6.1 are clear examples of vapor pressure isotherms indicating phase condensation. There are several studies on a variety of gas–substrate combinations, in which comparably convincing isotherms have been obtained, as well as others giving less clear but at least plausible indications. The earliest of these studies was by Ross and Clark (12), who studied the adsorption of methane, ethane, and Xe on cube crystals [presumably (100) faces] of sodium chloride. At relatively low temperatures the isotherms had very steep but not quite vertical midsections, and at higher T the lengths of the steep sections decreased monotonically and finally disappeared at a fairly well-defined temperature. This temperature was identified as $T_c(2D)$. Its value for all three gases was estimated as being comparable with but definitely lower than the value $0.5T_c(3D)$ predicted by simple theory [i.e., for $\nu = 6$ in Eq. (6.2.5)].

Since the pioneering work, there have been vapor pressure studies using substrates of quite different substances: various alkali halides, graphitized carbon black and exfoliated graphite, graphite preplated with heavier gases, and basal plane surfaces of layer compounds such as $NiCl_2$ (13–25). These substrates have usually been studied in combination with the heavier noble gases Kr and Xe; sometimes with other gases. A list of these vapor pressure studies, all of which yielded

Table 6.1   Adsorption Systems Displaying Phase Condensation in Vapor Pressure Isotherms

Substrate	Gas	Ref.
NaCl	Xe, CH_4, C_2H_6	12
NaBr	Kr	13
LiF	Kr	16
NaF	Kr	16
NaCl	Kr	17
KCl	Kr	17
RbCl	Kr	17
Graphite	Ar, O_2, N_2	14
Graphite	Kr, Xe, CH_4	19
Graphite	NO	20
Graphite	Ne	21
Graphite	Ar	22
Graphite + one Xe layer	Ar, O_2	14
Graphite + one Ar layer˙	Ar	14
Graphite + one Ne Layer	Ne	24
BN	NO	22
$NiCl_2$	Ar, Kr, Xe	18
$CoCl_2$	Ar, Kr, Xe	18
$FeCl_2$	Ar, Kr, Xe	18
$CdCl_2$	Ar, Kr, Xe	18
$CdBr_2$	Ar, Kr, Xe	18
CdI_2	Ar, Kr, Xe	18
PdI_2	Ar, Kr, Xe	18

isotherms having vertical or nearly vertical risers indicating phase condensation, is given in Table 6.1. In a few systems, notably those studied by Thomy and Duval (4, 19), there is a second vertical or nearly vertical portion occurring in most of the isotherms near the completion of the first layer. This feature, which is evident in Fig. 6.1, is identified with a 2D liquid–solid phase change. Because the nature of melting is entirely different from that of condensation we postpone discussion of these experimental features until Chapters 7 and 8 dealing with 2D solids and epitaxial phases.

Concerning the ratio $T_c(2D)/T_c(3D)$ (see Table 6.2), Thomy and Duval have obtained values very close to 0.4 for Ar, Kr, Xe, and CH_4, all on exfoliated graphite (19). A value somewhat lower for Ne on graphite was estimated by Steele and Karl from heat capacity measurements (23), but vapor pressure measurements by Huff gave $T_c(2D)/T_c(3D) = 0.45$ for the second layer of Ne on graphite (24). Lerner, Hegde, and Daunt (25) measured several vapor-pressure isotherms of one- and two-layer Ne films on graphite and their results also indicate adatom interactions in the second layer. Larher (18) found ratios of about 0.4 for Kr on cleavage planes of $CdCl_2$ and $NiCl_2$, and about 0.52 for Ar on $CdCl_2$. A number of suggestions have been made to explain the fact that the experimental ratio usually falls below the

Table 6.2 Two-Dimensional Critical Temperatures of Monolayers, Compared with the Critical Temperatures of the Adsorbates in Bulk

Gas	Substrate	$T_c(3D)$ (°K)	$T_c(2D)$ (°K)	$T_c(2D)/T_c(3D)$	Ref.
Ne	Graphite	44.4	16	0.36	23
Ne (2nd layer)	Graphite	—	20	0.45	24
Ar	Graphite	151	65	0.43	19
Ar	$CdCl_2$	—	78	0.52	18
Kr	Graphite	209	87	0.41_5	19
Kr	$CdCl_2$	—	90	0.43	18
Kr	$NiCl_2$	—	84	0.40	18
Xe	Graphite	290	117	0.40	19
CH_4	Graphite	191	75	0.39	19

value $\frac{1}{2}$ predicted from simple theory. Some explanations involve substrate-mediated interactions such as the effective repulsions due to induced dipoles, but these are not able to account for the insensitivity of the empirical ratio to the type of surface. In the early work of Ross and Clark on NaCl (12) and of Fisher and McMillan on NaBr (13) the ratios were found to be near 0.4. Larher (18) has argued that the ratio in fact does not have a universal value and that the amount by which it falls below 0.5 is a consequence of the mismatch between the substrate structure and the equilibrium spacing between atoms of the unstressed monolayer. The basic notion is simply that the unstressed state of the monolayer is the state of lowest free energy, and that the surface structure of an arbitrary adsorbing surface will most probably stretch or compress the layer to a less favorable configuration. Larher supports this idea by some quantitative arguments drawn from his studies on the layer halide surfaces.

An alternative possibility is that the effects are more general, having to do with the 2D van der Waals equation itself. The discrepancies between the experimental and theoretical values for $T_c(2D)$ are comparable in magnitude to the discrepancies in $T_c(3D)$, hence it is not surprising that the ratios $T_c(2D)/T_c(3D)$ should deviate from that of the approximate theory. Indeed, the van der Waals equation is derived on the basis of the high temperature, low density approximations (6.1.13) and (6.1.14), and they can not be justified for densities and temperatures near the critical point.

All of the experiments just discussed involve vapor pressure measurements: in a small number of other studies, monolayer condensation has been observed by other methods as well. The LEED survey by Lander and Morrison (26) of adsorption on the basal plane surface of graphite yielded indications of phase changes from ordered to liquid or amorphous solid phases in Br_2, Cs, and Xe monolayers. More recent LEED work on metal surfaces have shown a variety of abrupt transitions from disordered to regular structures: these systems include Xe on (100) Pd (27), CO on (100) Pd (28), and CO on (100) Ni and (100) Cu surfaces (29). Calorimetric measurements on adsorbed He, Ne and H_2 films have yielded evidence for transitions between gaseous, liquid, and solid phases. The gas–liquid transitions are discussed in Section 6.5 and the ordered and solid phases in Chapters 7 and 8.

6.4 TWO-DIMENSIONAL VIRIAL EXPANSION†

6.4.1 Series Expansion for Classical and Quantum Gases

The theory in Section 6.1 is a simple first approximation to the more general problem of an interacting monolayer, being based upon the independent assumptions of low density with respect to interatomic forces and also with respect to quantum statistical interactions. At higher densities additional correction terms become important, and the various thermodynamic quantities in principle can be evaluated as a series of terms involving the interactions of larger numbers of atoms. Ursell and Mayer (32) developed the general cluster expansion for classical gases with pair interactions and their method is directly applicable to monolayers, provided that the substrate is sufficiently smooth. Kahn and Uhlenbeck (33) generalized the cluster expansion to classical or quantum systems with arbitrary interaction potentials. A simpler general formalism was later given by Kilpatrick (34), which can be adapted to the problem of the interacting 2D gas, as follows.

The grand partition function of an interacting gas (independent of its dimensionality) can be written as

$$\varrho(z, A, T) = \sum_{l=0}^{\infty} Z_l(A, T) z^l, \qquad (6.4.1)$$

where $z = e^{\beta \mu}$ is the absolute activity and $Z_l(A, T)$ is the canonical partition function for l particles. If the logarithm of the grand partition function is expanded in a power series in z, called the cluster expansion,

$$(1/A) \log \varrho(z, A, T) = \sum_{l=1}^{\infty} b_l(A, T) z^l \qquad (6.4.2)$$

then the coefficient $b_l(A, T)$, called the lth cluster integral, can be found by comparing the series expansion of $\exp [\log \varrho(z, A, T)]$ with

---

† This discussion follows parts of Siddon and Schick (30) and the more extensive treatment of Siddon (31).

Eq. (6.4.1). The procedure is straightforward and the result is

$$b_l(A, T) = (1/A) \sum_{\{m\}} (-1)^{\delta-1}(\delta - 1)! \prod_{j=1}^{l} Z_j m_j(A, T)/m_j!, \quad (6.4.3)$$

where $\delta \equiv \sum_{j=1}^{l} m_j$ and the summation is over all sets of positive integers m_j including zero such that $\sum_{j=1}^{l} j m_j = l$.

The expansion of the logarithm of the grand partition function, Eq. (6.4.2), with $b_l(A, T)$ given by Eq. (6.4.3) is the principal result of the cluster expansion technique. The thermodynamic functions of the system are found by eliminating the quantity z from the parametric equations of state for the spreading pressure

$$\phi/kT = (1/A) \log \mathcal{Q}(z, A, T) = \sum_{l=1}^{\infty} b_l(A, T)z^l \quad (6.4.4)$$

and particle density

$$n = (1/A)z(\partial/\partial z) \log \mathcal{Q}(z, A, T) = \sum_{l=1}^{\infty} l b_l(A, T)z^l. \quad (6.4.5)$$

In particular, eliminating z by expressing the pressure in a power series in the density yields the virial expansion of the equation of state,

$$\phi/kT = \sum_{l=1}^{\infty} B_l(A, T)n^l, \quad (6.4.6)$$

where $B_l(A, T)$ is called the lth virial coefficient. If the density, given by Eq. (6.4.5), is substituted into the virial expansion and the resulting expression compared with Eq. (6.4.4), then the virial coefficient can be found by solving in succession the system of equations

$$b_l(A, T) = \sum_{\{m\}} B_\delta(A, T)\delta! \prod_{j=1}^{l} [jb_j(A, T)]^{m_j}/m_j!. \quad (6.4.7)$$

The virial coefficients are given in terms of the partition functions by eliminating the cluster integrals in the above expression using Eq. (6.4.3).

In the thermodynamic limit, the area and number of particles of the system become infinite while the density $n = N/A$ remains finite. As this is the case of interest, the virial coefficient is redefined by

$$B_l(T) = \lim_{A \to \infty} B_l(A, T) \qquad (6.4.8)$$

and is assumed to exist in the limit.

It follows from the expansion of the pressure equation (6.4.6), together with the fact that in the limit of zero density all thermodynamic functions must reduce to their noninteracting classical values (denoted below by the subscript zero), that for the interacting system, the Helmholtz free energy is

$$F/NkT = (F_0/NkT) + \sum_{l=1}^{\infty} B_{l+1}(T)(n^l/l). \qquad (6.4.9)$$

The other thermodynamic functions can be found from this result.

Entropy

$$S/Nk = (S_0/Nk) - (1 - \beta(d/d\beta)) \sum_{l=1}^{\infty} B_{l+1}(T)(n^l/l). \qquad (6.4.10)$$

Chemical potential

$$\mu/kT = (\mu_0/kT) + \sum_{l=1}^{\infty} B_{l+1}(T)[(l + 1)/l]n^l \qquad (6.4.11)$$

Energy

$$E/NkT = (E_0/NkT) + \beta(d/d\beta) \sum_{l=1}^{\infty} B_{l+1}(T)(n^l/l). \qquad (6.4.12)$$

Specific heat at constant area

$$C/Nk = C_0/Nk - \beta^2(d^2/d\beta^2) \sum_{l=1}^{\infty} B_{l+1}(T)(n^l/l). \qquad (6.4.13)$$

Classically, the evaluation of the virial coefficient involves calculation of the multiple integrals occurring in the canonical partition functions. However, the lth quantum mechanical virial coefficient, being related to the l-particle partition function, represents at least an l-body problem, so that for most problems, the quantum mechanical virial expansion is truncated after the second virial coefficient. The range of temperature and density for which the truncated expansion is adequate must then be investigated in detail. From Eqs. (6.4.3) and (6.4.7), the second virial coefficient $B(T)$ is

$$B(T) = - \lim_{A \to \infty} A[Z_2(A, T) - \tfrac{1}{2}Z_1{}^2(A, T)]/Z_1{}^2(A, T). \quad (6.4.14)$$

If the virial expansion is truncated after the second virial coefficient, then the free energy equation (6.4.9) will have the same form as the classical approximation equation (6.1.4); hence expressions (6.1.6)–(6.1.11) for thermodynamic quantities are also appropriate for quantum systems.

6.4.2   Quantum Mechanical Second Virial Coefficient

The quantum mechanical canonical partition function for N identical particles is

$$Z_N(A, T) = \int \cdots \int d\mathbf{r}_N \sum_\alpha \psi_\alpha{}^*(\mathbf{r}_N) \exp [-\beta H(\mathbf{p}_N, \mathbf{r}_N)]\psi_\alpha(\mathbf{r}_N),$$

$$(6.4.15)$$

where $H(\mathbf{p}_N, \mathbf{r}_N)$ is the system Hamiltonian and $\{\psi_\alpha(\mathbf{r}_N)\}$ is any complete orthonormal set of wavefunctions. For bosons (fermions), the summation is only over those states which are totally symmetric (antisymmetric) under exchange of particle coordinates, including spin. In the absence of spin, the single particle partition function $Z_1(A, T)$ is just A/λ^2. The second quantum virial coefficient for spinless particles can therefore be expressed as

$$B(T) = \lim_{A \to \infty} (1/2A) \iint [1 - 2\lambda^4 \sum_\alpha \psi_\alpha{}^*(1, 2) \, e^{-\beta H(1,2)}\psi_\alpha(1, 2)] \, d\mathbf{r}_1 \, d\mathbf{r}_2.$$

$$(6.4.16)$$

For convenience, the wavefunctions $\psi_\alpha(1, 2)$ are taken to be eigenfunctions of the two-body Hamiltonian with corresponding eigenvalues E_α. The development is simplified by transforming from the variables $\mathbf{r}_1$ and $\mathbf{r}_2$ to the center-of-mass $\mathbf{R} = (\mathbf{r}_1 + \mathbf{r}_2)/2$ and relative $\mathbf{r} = \mathbf{r}_1 - \mathbf{r}_2$ coordinates. Under such a transformation the two-body wavefunction can be factored into a plane wave, describing the motion of the center-of-mass, and a part $\psi_\mathbf{k}(\mathbf{r})$, describing the internal or relative motion of the two particles. The energy eigenvalue E_α is then a sum of the kinetic energy of the center-of-mass and a relative eigenvalue $E_\mathbf{k}$. The relative wave functions $\psi_\mathbf{k}(\mathbf{r})$ and eigenvalues $E_\mathbf{k}$ are determined from the relative Schrödinger equation

$$(-\hbar^2/2\mu)\nabla_r^2\psi_\mathbf{k}(\mathbf{r}) + V(\mathbf{r})\psi_\mathbf{k}(\mathbf{r}) = E_\mathbf{k}\psi_\mathbf{k}(\mathbf{r}), \qquad (6.4.17)$$

where μ is the reduced mass $m/2$. After summing over all values of the center-of-mass momentum, the virial coefficient equation (6.4.16) becomes

$$B(T) = \lim_{A \to \infty} \tfrac{1}{2} \int [1 - 4\lambda^2 \sum_\mathbf{k} |\psi_\mathbf{k}(\mathbf{r})|^2 \exp(-\beta E_\mathbf{k})]\, d\mathbf{r}. \quad (6.4.18)$$

In the absence of any two-body interaction potential, the virial coefficient is given by its ideal value $B^{(0)}(T)$, where

$$B^{(0)}(T) = \lim_{A \to \infty} \tfrac{1}{2} \int [1 - 4\lambda^2 \sum_\mathbf{k} |\psi_\mathbf{k}^{(0)}(\mathbf{r})|^2 \exp -\beta E_\mathbf{k}^{(0)}]\, d\mathbf{r}. \quad (6.4.19)$$

The ideal wavefunctions $\psi_\mathbf{k}^{(0)}(\mathbf{r})$ and eigenvalues $E_\mathbf{k}^{(0)}$ satisfy Eq. (6.4.17) with $V(\mathbf{r})$ set to zero:

$$(-\hbar^2/2\mu)\,\nabla_r^2\psi_\mathbf{k}^{(0)}(\mathbf{r}) = E_\mathbf{k}\psi_\mathbf{k}^{(0)}(\mathbf{r}). \qquad (6.4.20)$$

By subtracting Eqs. (6.4.19) and (6.4.18) and using the fact that the relative wavefunctions for the interacting and noninteracting systems are normalized, we find the second virial coefficient reduces to

$$B(T) = B^{(0)}(T) + \lim_{A \to \infty} 2\lambda^2 \sum_\mathbf{k} [\exp(-\beta E_\mathbf{k}^{(0)}) - \exp(-\beta E_\mathbf{k})]. \quad (6.4.21)$$

The quantities in the sum of Eq. (6.4.21) are functions of the 2D wavevector $\mathbf{k}$. Because of the cylindrical symmetry of the potential the relative quantum numbers can be taken to be the usual azimuthal quantum number m and a radial quantum number k which is related to the relative energy according to

$$E_k^{(0)} = \hbar^2 k^2/2\mu, \qquad (6.4.22)$$

where μ is the reduced mass. The resulting sums over m contain only even positive and negative integers for bosons and odd positive and negative integers for fermions. Equation (6.4.21) is further simplified by writing the sum over bound states explicitly and integrating over the quasi-continuum using a density of states weighting factor. The resulting expression is

$$B(T) = B^{(0)}(T) - 2\lambda^2 \sum_B{}' \exp(-\beta E_B)$$

$$- 2\lambda^2 \int_0^\infty \Sigma'[g_m(k) - g_m^{(0)}(k)] \exp(-\beta\hbar^2 k^2/2\mu)\, dk, \qquad (6.4.23)$$

where the subscript B stands for bound states and the prime on the sum denotes a restriction to even or odd values of m.

The difference in density of states is related to the scattering phase shifts as follows. The relative wavefunction $\psi_{m,k}(r)$ can be factored into a product of an azimuthal part and a radial part $R_{m,k}(r)$. The radial wavefunction at large r, where the potential is assumed negligible, satisfies

$$R_{m,k}(r) \sim \cos[kr - \tfrac{1}{4}\pi - \tfrac{1}{2}m\pi + \delta_m(k)], \qquad (6.4.24)$$

which defines the phase shift $\delta_m(k)$ of the mth partial wave. The allowed values of k are determined by demanding that the wave function vanish on a circle of radius R which, at the end of the calculation, will be taken to infinity. This is equivalent to the statement for the non-interacting system,

$$\cos(kR - \tfrac{1}{4}\pi - \tfrac{1}{2}m\pi) = 0, \qquad m = 0, \pm 1, \pm 2, \ldots$$

Therefore, the allowed k values are determined from

$$kR - \tfrac{1}{4}\pi - \tfrac{1}{2}m\pi = (n + \tfrac{1}{2})\pi, \qquad n = 0, 1, 2, \ldots$$

The interacting system in general has a quasi continuum plus a discrete spectrum of negative or bound state energies E_B. The vanishing of the relative wavefunction for the interacting system is given by the condition

$$\cos\left[kR - \tfrac{1}{4}\pi - \tfrac{1}{2}m\pi + \delta_m(k)\right] = 0, \qquad m = 0, \pm1, \pm2, \ldots\ldots$$

In the asymptotic limit, the sole effect of the two-body potential is simply to shift the phase of the noninteracting partial wave by an amount $\delta_m(k)$. The allowed k values for the interacting system are determined from

$$kR - \tfrac{1}{4}\pi - \tfrac{1}{2}m\pi + \delta_m(k) = (n + \tfrac{1}{2})\pi, \qquad n = 0, 1, 2, \ldots\ldots \quad (6.4.25)$$

Let $g_m^{(0)}(k)$ denote the number of states with relative wave vector between k and $k + \Delta k$ for the mth partial wave of the noninteracting system and let $g_m(k)$ denote the corresponding quantity for the interacting system. The second virial coefficient, Eq. (6.4.22), can then be written as

$$B(T) = B^{(0)}(T) - 2\lambda^2 \sum_B \exp\left(-\beta E_B\right) - 2\lambda^2 \int_0^\infty \sum_m \left(g_m(k)\right.$$

$$\left. - g_m^{(0)}(k)\right) \exp\left(-\frac{\beta\hbar^2 k^2}{2\mu}\right) dk, \qquad (6.4.26)$$

where the summations are over bound states and density of states for m even $(0, \pm2, \pm4, \ldots)$ for bosons and m odd $(\pm1, \pm3, \ldots)$ for fermions. For the noninteracting system, the interval between allowed k values for the mth partial wave is π/R. Since there is only one state in each interval, the density of states $g_m^{(0)}(k)$ for the mth wave is simply R/π. For the interacting system, the corresponding density of states $g_m(k)$ is $(R + \partial\delta_m(k)/\partial k)/\pi$. The difference between the interacting and noninteracting density of states for the mth partial wave

$$g_m(k) - g_m^{(0)}(k) = (1/\pi)(\partial/\partial k)\delta_m(k) \qquad (6.4.27)$$

depends only on the variation of the corresponding phase shift with energy. The second virial coefficient can now be written as

$$B(T) = B^{(0)}(T) - 2\lambda^2 \sum_B \exp\left[-\beta E_B\right]$$

$$- (2\lambda^2/\pi) \int_0^\infty \sum_m (\partial/\partial k)\delta_m(k) \exp\left(-\beta\hbar^2 k^2/2\mu\right) dk, \qquad (6.4.28)$$

where the summation restrictions are the same as for Eq. (6.4.26). An equivalent form is obtained by an integration by parts and using the fact that at zero energy the phase shift $\delta_m(0)$ is equal to the number of bound states of azimuthal quantum number m times π. The result is

$$B(T) = B^{(0)}(T) - 2\lambda^2 \sum_B \left(\exp\left(-\beta E_B \right) - 1 \right)$$

$$- 2\lambda^4/\pi^2 \int_0^\infty k \sum_m \delta_m(k) \exp\left(-\beta \hbar^2 k^2/2\mu \right) dk. \quad (6.4.29)$$

The ideal second virial coefficient $B^{(0)}(T)$ is easily found from Eq. (6.4.20), using properly symmetrized plane waves. The result is

$$B^{(0)}(T) = \mp \lambda^2/4, \quad (6.4.30)$$

where the minus sign is for bosons.

The inclusion of spin in the virial coefficient formalism is made by explicitly using the canonical partition functions in Eq. (6.4.3) for a system of identical particles with spin. This is particularly simple for the second virial coefficient as it involves only $Z_1(A, T)$ and $Z_2(A, T)$. The procedure is easily generalized to higher coefficients.

For one particle with spin s, every state is degenerate by the factor $2s + 1$. Therefore, the one-particle canonical partition function for spin s is just $2s + 1$ times the partition function for no spin;

$$Z_1^{(s)}(A, T) = (2s + 1)Z_1(A, T). \quad (6.4.31)$$

For a system of two identical particles each with spin s, there are $(2s + 1)^2$ total spin states. Of the $(2s + 1)^2$ states, $(s + 1)(2s + 1)$ are symmetric with respect to exchange of spin coordinates and $s(2s + 1)$ are antisymmetric. It follows then, since the two-particle partition function is a sum over all symmetric or antisymmetric states, that for bosons (BE) with spin s,

$$Z_2^{(s)\mathrm{BE}}(A, T) = (s + 1)(2s + 1)Z_2^{\mathrm{BE}}(A, T) + s(2s + 1)Z_2^{\mathrm{FD}}(A, T) \quad (6.4.32)$$

and for fermions (FD) with spin s,

$$Z_2^{(s)\mathrm{FD}}(A, T) = (s + 1)(2s + 1)Z_2^{\mathrm{FD}}(A, T) + s(2s + 1)Z_2^{\mathrm{BE}}(A, T), \quad (6.4.33)$$

where the partition functions on the right are for no spin.

The second virial coefficient for a system with spin s is given by Eq. (6.4.14), using the appropriate partition functions $Z_1^{(s)}$ and $Z_2^{(s)}$ according to the statistics. Thus, the virial coefficient for bosons with spin s is

$$B_{\mathrm{BE}}^{(s)}(T) = [(s+1)/(2s+1)]B_{\mathrm{BE}}(T) + [s/(2s+1)]B_{\mathrm{FD}}(T),$$
(6.4.34)

where the quantities on the right are the virial coefficients for no spin. The corresponding quantity for fermions with spin s follows in the same manner, leading to the result

$$B_{\mathrm{FD}}^{(s)}(T) = [(s+1)/(2s+1)]B_{\mathrm{FD}}(T) + [s/(2s+1)]B_{\mathrm{BE}}(T).$$
(6.4.35)

For systems with distinguishable (Boltzmann) particles, the N particle canonical partition function is $1/N!$ times the partition function calculated using all the states for N particles. Therefore, the two-particle partition function is

$$Z_2(A, T) = \tfrac{1}{2}(Z_2^{\mathrm{BE}}(A, T) + Z_2^{\mathrm{FD}}(A, T)).$$
(6.4.36)

The second virial coefficient for Boltzmann particles is obtained by substituting this into Eq. (6.4.14). The result is

$$B(T) = \tfrac{1}{2}(B_{\mathrm{BE}}(T) + B_{\mathrm{FD}}(T)).$$
(6.4.37)

It is evident from Eqs. (6.4.34) and (6.4.35) that the Boltzmann coefficient is the boson or fermion coefficient in the limit of infinite spin.

6.5 VIRIAL COEFFICIENTS OF MONO-LAYER FILMS; EXPERIMENT AND THEORY

6.5.1 Argon

Several experimental measurements have been made of the 2D virial coefficients of adsorbed films. Sams et al. (35) studied the vapor pressure isotherms of Ar on graphitized carbon black at temperatures 140–240°K and a range of relatively low coverages. They obtained qualitatively good agreement between the measured second virial coefficients and those calculated for a classical 2D gas having a 6–12 pair potential with the empirical Ar gas parameters. The temperature dependence of the 2D B was found to be in reasonable accord with the calculation, and changed sign at very nearly the correct computed "2D Boyle" temperature. Quantitatively, however, the agreement was found to be poor. Over a factor of two, discrepancy in the magnitude of B could be accounted for by a reduction in the depth of the potential of about 20%. A reduction of this magnitude might be caused by a substrate mediated interaction of the induced static or fluctuating dipole variety (see Chapter 2). An error in the calculations was subsequently discovered and several corrected tables of the classical 2D B were published (36–40). The most extensive calculation is that of Morrison and Ross (40), who calculated terms up to the sixth virial coefficient for a classical 2D 6–12 gas with parameters corresponding to bulk argon. They found that the experimental isotherms (35) could be completely described by their computed series.

6.5.2 ^{3}He and ^{4}He

Lighter gases have presented greater problems, for the region of validity of the classical approximation is at higher temperature. In ^{4}He, for example, appreciable quantum corrections begin at temperatures on the order of 100°K. Since there is virtually no adsorption of He at such high temperatures there is no experimental regime in which the classical approximation applies. Steele and Derderian (41) extended the theoretical treatment to lower temperatures by a perturbation expansion in powers of $\hbar$, carrying out the calculations as far as the quadratic term. They computed values for B for 2D 6–12 gases with parameters corresponding to bulk He and compared these with vapor pressure isotherms of ^{3}He and ^{4}He on graphitized carbon black between 10 and 70°K. They found only qualitative agreement in the magnitudes of B, and major discrepancies in the magnitudes and sign of dB/dT. The discrepancies were attributed by the authors to the neglect of the substrate structure. However, subsequent calculations by Siddon and Schick (30, 31, 42) indicate that the inadequacy of the perturbation expansion over the experimental range could have been the major factor in the discrepancy.

Siddon and Schick carried out extensive numerical computations of the fully quantum mechanical, second virial coefficients of 2D gases, according to the theory outlined in the previous section, in various approximations to ^{3}He and ^{4}He. They studied hard disks with and without statistical effects with different masses, Lennard-Jones 6-12 molecules with parameters appropriate to the bulk gases (30, 31) and the same molecules fitted with a Beck potential (42). The computed coefficients and their effects on heat capacities were compared with the calorimetric measurements taken at the University of Washington, of ^{3}He and ^{4}He on graphite (43–45) and of the calorimetric and vapor pressure studies at California Institute of Technology, of ^{4}He on graphite (46). The theoretical work was initiated by the experimental indications that the He films at low coverage were basically acting as 2D mobile gases, with isotope-specific interactions increasing in importance at temperatures below 1 or 2°K. Figures 6.2 and 6.3 illustrate the results for ^{3}He and ^{4}He at temperatures above 1°K, and Fig. 6.4 shows the ^{3}He results at very low temperatures. The signals are relatively constant at $C/Nk \simeq 1$ for both isotopes at the higher temperatures, but deviations of opposite sign begin near 2°K. [A proposed

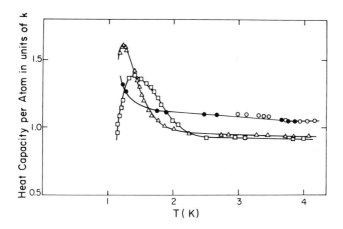

FIG. 6.2 Specific heats of low density ^{4}He films adsorbed on Grafoil (45). Densities in (Å)$^{-2}$ are: (○) 0.0158; (●) 0.0164; (△) 0.0273; (□) 0.040. Completed monolayer density of ^{4}He is approximately 0.115 (Å)$^{-2}$.

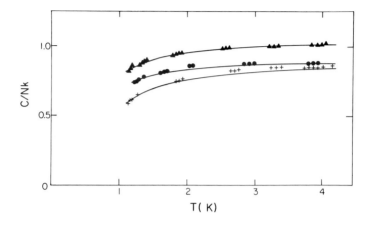

FIG. 6.3 Specific heats of low density ^{3}He films adsorbed on Grafoil (45). Densities in (A)$^{-2}$ are: (▲) 0.0154; (●) 0.0273; (+) 0.0415. Completed mono-layer density of ^{3}He is approximately 0.11 (Å)$^{-2}$. The smooth curves are a guide to the eye.

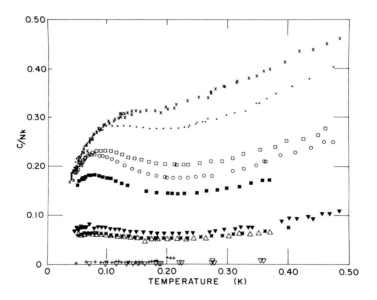

FIG. 6.4 Specific heats of low density and low temperature ^{3}He films adsorbed on Grafoil (44,45). Densities in (Å)$^{-2}$ are: (×) 0.0279; (●) 0.0332; (□) 0.0420; (○) 0.0452; (■) 0.0485; (▼) 0.0548; (*) 0.0564; (△) 0.0571; (+) 0.0662; (▽) 0.0694.

explanation for the deviations was given in terms of substrate heterogeneities (see Chapter 9, Section 9.5) but in view of the success of the
calculated virial corrections, it was concluded that heterogeneity did
not need to be invoked.]

The deviations were explained as due to interactions between the gas
atoms adsorbed on the surface. For both isotopes there was found to be
substantial quantitative agreement between $(C/Nk - 1)$ and the virial
corrections for He interacting by either Lennard-Jones or Beck potentials. Hard-disk potentials gave good agreement with the ^{4}He data,
both in the magnitude and temperature dependence of the specific heat
correction. However the hard-disk potential could not be made to give
even qualitative agreement with ^{3}He results, for it gives a strong rise
in $\beta^2 \, d^2B/d\beta^2$ for both bosons and fermions at low temperatures. The
distinction illuminates the behavior of the actual films, the decrease in
the ^{3}He specific heat at low T evidently coming from the attractive
part of the potential. ^{4}He, on the other hand, is apparently more
sensitive to the short-range repulsive part of the potential, which it can
sample more effectively by its ground state s-wave relative wavefunction. Figures 6.5 and 6.6 illustrate the comparisons between the
calculated and experimental results, plotted in a sensitive manner
designed to emphasize the deviations from ideality. The virial expansion for the heat capacity [Eq. (6.4.13)] when truncated after the
second term can be written in the form

$$n^{-1}(C/Nk - 1) = -\beta^2(d^2B/d\beta^2). \qquad (6.5.1)$$

Therefore, the adequacy of the truncated virial expansion can be
tested by plotting the experimental results as $n^{-1}(C/Nk - 1)$ versus
T, to test whether it is independent of the film density n. If this test is
satisfactory, then a comparison can be made with computed coefficients, i.e., the right-hand side of Eq. (6.5.1). As shown in Figs. 6.5
and 6.6, both tests are well satisfied, although the region of validity
for ^{4}He is much more restricted than that of ^{3}He. For ^{3}He the range
extends from about 0.4 to at least 4°K, the upper limit of the experimental study. Over most of this range, the computed correction is in
very close quantitative agreement with the measurements. At temperatures lower than 1°K the data and the calculations diverge systematically, yet the data remain independent of n until a much lower temperature. This indicates a region in which the truncated virial expansion

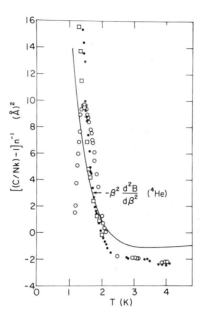

FIG. 6.5 Comparison of calculated (31) and observed (45) specific heat deviation per unit density for ^{4}He. Densities in $(\text{Å})^{-2}$ are: ($\square$) 0.0268; ($\bullet$) 0.0273; ($\bigcirc$) 0.0399.

continues to be valid but the computed correction is not correct. The divergence can come about through an error in the effective pair potential, which might be due to inherent limitations in the 6-12 shape, or through changes in the parameters due to substrate mediated interactions. Siddon and Schick explored the question by recomputing the virial coefficients using a Beck potential fitted to bulk He (42). The Beck potential differs from the Lennard-Jones in that the repulsion is somewhat "softer" and the minimum occurs at a separation which is 0.1 Å larger. The change to the Beck potential did not significantly modify the discrepancies between the computed and measured specific heat corrections. These negative findings suggest that the pair potential is appreciably modified by substrate effects.

The ^{4}He results are qualitatively different from ^{3}He, showing substantial specific heat peaks near 1°K. The calculated correction rises rapidly in this region. Both experiment and theory are believed to

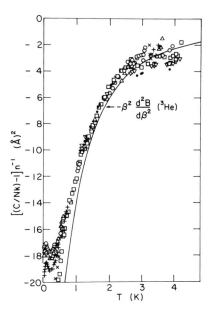

FIG. 6.6 Comparison of calculated (31) and observed (45) specific heat deviation per unit density for ^{3}He. Densities in $(\text{Å})^{-2}$ are: ($\bullet$) 0.0274; ($\square$) 0.0279; ($\times$) 0.0330; (∇) 0.0415; ($+$) 0.0417; ($\triangle$) 0.0451; ($\bigcirc$) 0.0483.

indicate 2D liquefaction. Siddon and Schick showed that their calculated B for ^{4}He could be closely approximated by the van der Waals form, i.e.,

$$B = b - \beta a$$

over a substantial region (1–10°K) covering the range of interest. The corresponding values of a and b yielded the critical parameters [see Eq. (6.2.3)] $T_c = 1.93°$K and $n_c = 0.0437$ $(\text{Å})^{-2}$. The experimental peaks, if interpreted as condensation, indicate approximate values $T_c = 1.4°$K, $n_c = 0.04$ $(\text{Å})^{-2}$. Similar procedures for bulk He led to quite comparable shifts between calculated and experimental critical temperatures and densities, therefore the critical parameters deduced from the calculated B are consistent with the interpretation of the peaks as due to liquefaction. A problem arises, however, in fitting the peaks to the proper theoretical shapes. The specific heat of an evaporating system should show the qualitative behavior of Fig. 4.4 (Chapter

4), having a continuous rise with increasing T up to the temperature at which the last drop of liquid disappears, when there is a discontinuous decrease. The experimental peaks, on the other hand, are definitely rounded and show no abrupt or steep changes. The rounded peaks may be due to a superposition of ideal sharp peaks located at slightly different temperatures. A range of evaporation temperatures could be caused by a certain metastable density variation throughout the sample. For a semiquantitative discussion of such an effect see Chapter 9, Section 9.6. Still another piece of evidence for liquefaction is the good agreement found between the experimental low coverage ^{4}He results below 0.5°K and the theoretical properties of 2D ^{4}He liquid at very low temperatures (47), both in the quadratic dependence on temperature and the numerical values of the coefficient.†

In contrast to the situation in ^{4}He films, there is no experimental evidence for liquefaction of ^{3}He monolayers at any temperature. The lack of condensation is consistent with theoretical expectations that the ^{3}He system is more weakly bound than ^{4}He (49–52). It is possible that the n_c of He^3 is lower than the surveyed range, which could account for the absence of condensation peaks in the experimental heat capacity.

6.5.3 Hydrogen

Bretz and Chung (53) measured the heat capacities of para-H_2 films adsorbed on graphite at several coverages below one monolayer, and at temperatures from 1 to 20°K. They observed rounded peaks near 12°K resembling the anomalies seen in ^{4}He films, suggesting a similar mechanism of 2D condensation. The high temperature sides of the peaks are in reasonable agreement with the density dependence of the truncated series equation (6.4.13). Siddon and Schick (54) calculated the second virial correction appropriate to this system and deduced a critical temperature of about 8.5°K. The agreement was considered satisfactory in view of the nature of the van der Waals approximation. A particularly interesting result of the experiments is the absence of

† The authors subsequently discovered an error of a factor of four missing in the calculated values for specific heat in Ref. 47. When this is corrected there is close agreement with the experiments (48).

any indication of 2D solidification down to temperatures as low as 1°K. This leaves open the possibility, originally suggested by Ginzburg and Sobyanin (55), that if solidification in H_2 films could be postponed to temperatures below 6°K they might display superfluidity.

REFERENCES

1. N. K. Adam, "The Physics and Chemistry of Surfaces," Oxford Univ. Press (Clarendon), London and New York, 1941.

2. N. L. Gershfeld, "Techniques of Surface and Colloid Chemistry and Physics" (R. J. Good, ed.), Vol. 1, p. 1.Dekker, New York, 1972.

3. G. A. Hawkins and G. B. Benedek, *Phys. Rev. Lett.* **32**, 524 (1974).

4. A. Thomy and X. Duval, *J. Chim. Phys. Physicochim. Biol.* **67**, 1101 (1970).

5. M. Volmer, *Z. Phys. Chem.* **115**, 253 (1925).

6. R. Fowler and E. A. Guggenheim, "Statistical Thermodynamics." Cambridge Univ. Press, London and New York, 1939.

7. H. M. Cassel, *J. Phys. Chem.* **38**, 195 (1944).

8. T. L. Hill, *J. Chem. Phys.* **14**, 441 (1946); **17**, 520 (1949).

9. W. A. Steele and G. D. Halsey, Jr., *J. Chem. Phys.* **22**, 979 (1954).

10. W. A. Steele, *in* "The Solid-Gas Interface," (E. A. Flood, ed.), Vol. 1, p. 307. Dekker, New York, 1967.

11. L. Landau and E. M. Lifshitz, "Statistical Physics." Pergamon, Oxford and Addison-Wesley, Reading, Massachusetts, 1958.

12. S. Ross and H. Clark, *J. Amer. Chem. Soc.* **76**, 4291, 4297 (1954).

13. B. B. Fisher and W. G. McMillan, *J. Amer. Chem. Soc.* **79**, 2969 (1957); *J. Chem. Phys.* **28**, 549, 555, 562 (1958).

14. C. F. Prenzlow and G. D. Halsey, Jr., *J. Phys. Chem.* **61**, 1158 (1957).

15. X. Duval and A. Thomy, *C. R. Acad. Sci. Paris* **259**, 4007 (1964).

16. Y.-F. YuYao, *J. Phys. Chem.* **69**, 2472 (1965).

17. T. Takaishi and M. Saito, *J. Phys. Chem.* **71**, 453 (1967).

18. Y. Larher, *J. Phys. Chem.* **72**, 1847 (1968); *J. Chim. Phys. Physicochim. Biol.* **68**, 796 (1971); *J. Colloid Interface Sci.* **37**, 836 (1971).

19. A. Thomy and X. Duval, *J. Chim. Phys. Physicochim. Biol.* **66**, 1966 (1969); **67**, 286, 1101 (1970).

20. M. Matecki, A. Thomy, and X. Duval, *C. R. Acad. Sci. Paris* **273**, 1485 (1971).

21. A. Thomy, X. Duval, and J. Regnier, *C. R. Acad. Sci. Paris* **268**, 1416 (1969).

22. M. Matecki, A. Thomy, and X. Duval, *C. R. Acad. Sci. Paris* **274**, 15 (1972).

23. W. A. Steele and R. Karl, *J. Colloid Interface Sci.* **28**, 397 (1968).

24. G. B. Huff and J. G. Dash, *Low-Temp. Phys.—LT* **13**, 1; G. B. Huff Ph. D. Thesis, Univ. of Washington, 1972.

25. E. Lerner, S. G. Hegde, and J. G. Daunt, *Phys. Lett.* **41A**, 239 (1972).

26. J. J. Lander and J. Morrison, *Surface Sci.* **6**, 1 (1967).

27. P. W. Palmberg, *Surface Sci.* **25**, 598 (1971).

28. J. C. Tracy and P. W. Palmberg, *J. Chem. Phys.* **51**, 4852 (1969).

29. J. C. Tracy, *J. Chem. Phys.* **56**, 2736, 2748 (1972).

30. R. L. Siddon and M. Shick, *Phys. Rev.* **A9**, 907 (1974).

31. R. L. Siddon, Ph.D. Thesis, Univ. of Washington, 1973 (unpublished).

32. See J. E. Mayer and M. G. Mayer, "Statistical Mechanics." Wiley, New York, 1940.

33. B. Kahn and G. E. Uhlenbeck, *Physica* **5**, 399 (1938).

34. J. Kilpatrick, *J. Chem. Phys.* **21**, 274 (1953).

35. J. R. Sams, G. Constabaris, and G. D. Halsey, Jr., *J. Chem. Phys.* **36**, 1334 (1962).

36. J. A. Barker and D. H. Everett, *Trans. Faraday Soc.* **58**, 1608 (1962).

37. J. D. Johnson and M. L. Klein, *Trans. Faraday Soc.* **60**, 1964 (1964).

38. R. Wolfe and J. R. Sams, *J. Chem. Phys.* **44**, 2181 (1966).

39. T. B. MacRury and J. R. Sams, *J. Chem. Phys.* **51**, 1302 (1969).

40. I. D. Morrison and S. Ross, *Surface Sci.* **39**, 21 (1973).

41. W. A. Steele and E. J. Derderian, in "Adsorption–Desorption Phenomena," (F. Ricca, ed.). Academic Press, New York, 1972.

42. R. L. Siddon and M. Schick, *Phys. Rev.* **A9**, 1753 (1974).

43. M. Bretz and J. G. Dash, *Phys. Rev. Lett.* **26**, 963 (1971).

44. D. C. Hickernell, E. O. McLean, and O. E. Vilches, *Phys. Rev. Lett.* **28**, 789 (1972).

45. M. Bretz, J. G. Dash, D. C. Hickernell, E. O. McLean, and O. E. Vilches, *Phys. Rev.* **A8**, 1589 (1973); **A9**, 2814 (1974).

46. R. L. Elgin and D. L. Goodstein, *Phys. Rev.* **A9**, 2657 (1974).

47. M. D. Miller and C.-W. Woo, *Phys. Rev.* **A7**, 1322 (1973).

48. C.-W. Woo, private communication.

49. R. H. Anderson and T. C. Foster, *Phys. Rev.* **151**, 190 (1966).

50. C. E. Campbell and M. Schick, *Phys. Rev.* **A3**, 691 (1971).

51. A. Bagchi, *Phys. Rev.* **A3**, 1133 (1971).

52. M. D. Miller, C.-W. Woo, and C. E. Campbell, *Phys. Rev.* **A6**, 1942 (1972).

53. M. Bretz and T. T. Chung, *J. Low-Temp. Phys.* **17**, 479 (1974).

54. R. L. Siddon and M. Schick, *J. Low-Temp. Phys.* **17**, 489 (1974).

55. V. L. Ginzburg and A. A. Sobyanin, *Sov. Phys.-JETP Lett.* **15**, 242 (1972).

7. Solid Phases and Melting Phenomena

The high density regimes of many monolayer and multilayer systems display a variety of characteristics that can be termed "solid-like." In this class one can discern two main divisions. The first, which is discussed in the present chapter, can be understood (to a first approximation, at least) in terms of 2D solid models, where the substrate appears to act primarily as a smooth adsorbing plane. The second group, treated in the next chapter, involves epitaxial films in which the substrate structure evidently plays an important role.

7.1 SURVEY OF EXPERIMENTAL RESULTS

7.1.1 ^{3}He and ^{4}He on Graphite

Two-dimensional solid behavior in monolayers of ^{3}He and ^{4}He adsorbed on graphite has been observed by calorimetric, vapor pressure, and nuclear resonance techniques (1–10). The characteristics of these systems correspond closely with theoretical 2D solid models and with semiquantitative scaling of the properties of 3D solid He.† The ob-

---

† Several previous studies of the heat capacities of ^{3}He and ^{4}He adsorbed on other substrates (jeweler's rouge, porous Vycor glass, sintered copper, and the

servations are reviewed below, and further discussion of the data is
given in connection with the theory in Sections 7.2 and 7.3.

One of the most distinctive characteristics of theoretical 2D solids
is the low temperature behavior of the heat capacity, varying as T^2 as
$T \to 0$. In the ³He and ⁴He monolayers on graphite (11) at densities
$n \lesssim 0.08$ (Å)$^{-2}$ the heat capacities vary as T^2 up to relative temperatures $T/\theta(2D) \simeq 0.07$ (12). $\theta(2D)$ is the "2D Debye" temperature,
which is proved in Section 7.2 to be suitable for characterizing the low
temperature specific heat of arbitrary 2D solids composed of particles
interacting by typical short-range forces. The range of constancy of the
empirical $\theta(2D)$ of ⁴He is comparable to that of the 3D $\theta(3D)$ of the
corresponding bulk solid. A very strong dependence of the heat capacity
on coverage was observed. As can be seen in Fig. 7.1, the empirical
$\theta(2D)$ for ⁴He changes by over a factor of two for an increase of 30% in
areal density. This sensitivity of the characteristic temperature to
surface density change was suggestive of the strong dependence of
$\theta(3D)$ on bulk density, and to effect a more quantitative comparison
between the films and bulk Bretz et al. (1, 3) used a simple scaling
scheme. They placed both the film data and Ahlers' measurements of
hcp ⁴He (13) on a common area scale, defining the effective molecular
area of the bulk solids as $v^{2/3}$, v being the molecular volume. With this
scaling scheme, illustrated in Fig. 7.1, they found that the data for
$\theta(2D)$ and $\theta(3D)$ are virtually identical over the entire range common
to both studies. (The quantitative scale for areal density was established by an adsorption area calibration based on the $x_g = \frac{1}{3}$ epitaxial
phase of ⁴He described in Chapter 8.) It was perhaps this agreement
more than any other single feature of the initial results that provided a
convincing indication of 2D solid behavior. Since that initial report,

---

latter two with a variety of physisorbed gases as preplating) showed T^2 heat
capacities at low temperatures. For several years it had been felt that this dependence indicated 2D solid behavior. However, a series of experiments undertaken to
explore the variation of $\theta(2D)$ with film coverage failed to disclose any effects of
2D melting or sublimation and led to an anomalously high latent heat of lateral
binding. These results, together with the qualitatively different and more reasonable behavior of films adsorbed on graphite are taken as evidence of strong
lateral fields due to binding energy variations on the previous substrates. For a
more detailed discussion of the earlier work, see Ref. 3, and for a discussion of
heterogeneity, see Chapters 3 and 9.

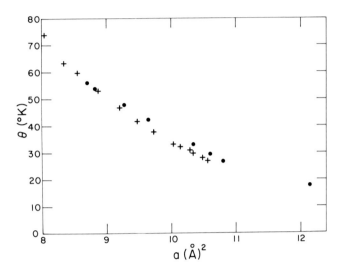

FIG. 7.1   2D and 3D Debye temperatures of ⁴He monolayers on graphite (1, 3)
(●) and hcp ⁴He (11) (+) on a common molecular-area scale.

however, there have been more detailed and sophisticated analyses
(see Section 7.2.3) pointing out that the simple $\frac{2}{3}$ scaling law can be
only a semiquantitative gauge, and, therefore, the fact that the curves
are superimposed in Fig. 7.1 is not to be taken as fundamentally
significant.

Also observed in the initial experiments were a series of pronounced
heat capacity peaks which, by qualitative arguments and a scaling
comparison with bulk ⁴He, were identified as due to melting. These
peaks, shown in Fig. 7.2, shift to higher temperatures with increasing
coverage and become impressively high and sharp near full monolayer
coverage. The peak temperatures are comparable in value to the melting temperatures of hcp solid ⁴He on a common scale of effective
molecular area, as seen in Fig. 7.3. The similarities in values and
trends were the basis for the melting identification, although the
qualitative differences between the shapes of the peaks and the characteristic shapes of first-order phase transitions made the assignment
somewhat tentative.

Strong support for the melting identification came within a year

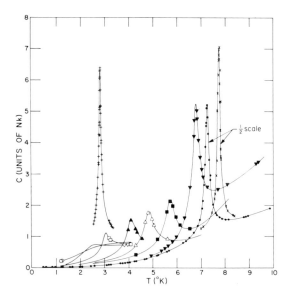

FIG. 7.2 High density ⁴He monolayer "melting anomalies" (1, 3). Monolayer densities n Å⁻² are: (○) 0.075; (▽) 0.0873; (□) 0.0942; (▲) 0.0967; (△) 0.0991; (■) 0.104; (▼) 0.108; (●) 0.115; (×) 0.133. Completed first layer coverage is approximately 0.115 (Å)⁻².

from Rollefson's study of nuclear magnetic resonance of ³He on graphitized carbon black (2). In this experiment the resonance line, which has a width of about 1 G at relatively low temperatures and densities approaching a completed monolayer, was found to narrow rapidly to about one-third of that value within a small temperature range near 3°K. The broad line at low temperature was calculated to have a width consistent with static dipolar broadening due to nearest neighbors at the density of the 2D solid phase; the decrease in width on warming was attributed to motional narrowing, i.e., due to the increase in mobility on melting. The narrowing occurred at densities and temperatures consistent with the ⁴He "melting line" determined from the heat capacity peaks, but at the time of the nmr study there had not yet been a similar heat capacity study of ³He. Since then, the melting line of ³He has been delineated by Hering and Vilches (4, 10) in the same manner as for ⁴He, and it has indeed been found to lie quite close to

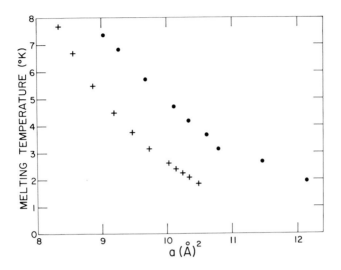

FIG. 7.3   Temperatures of the specific heat anomalies in ⁴He monolayers (1, 3)
(●) and melting temperatures of hcp ⁴He (11) (+) on a common molecular-
area scale.

the ⁴He curve. Thus the nmr study has borne out the original conclusion that the strong heat capacity peaks are due to some sort of a
solid–fluid transformation in the films (for a discussion of the peak
shapes and theoretical implications, see Section 7.3). Rollefson's study
is most unusual in the entire field of adsorption, in the extent to which
the nmr observations are correlated with other measurements.† This
correlation extends beyond the solid phase and its melting line, and
includes the entirely distinct lattice gas order-disorder transformation,
which is discussed at length in the next chapter.

In addition to the calorimetric and nmr studies there have been
careful vapor pressure measurements on high density monolayers of
He on similar graphite substrates. These have given considerably more
insight into the properties of the 2D solid phase and the nature of its
melting process, not only of He films on graphite but, quite generally,

† Independent studies of nmr in ⁴He films, using pulse techniques and other
substrates, have not yielded evidence for phase changes comparable with Rollefson's findings. For discussions of other work, see the reports by Brewer and
Grimmer (12).

of monolayers physisorbed on smooth surfaces. Stewart *et al.* (9) measured vapor pressures of ⁴He in the density range near monolayer completion, obtaining the isosteric heat of adsorption and chemical potential as a function of coverage at 4.2°K. In the region of completion, the vapor pressure was found to rise exponentially with coverage. This behavior is in approximate agreement with quantitative calculations, discussed in Section 7.2.3. Stewart, *et al.* also presented a theoretical analysis whereby their vapor pressure data could be combined with the specific heat measurements so as to yield experimental values for the phase velocities of the elastic waves in the film; i.e., for the transverse and longitudinal branches independently. Subsequently Stewart extended these ideas and the detailed analysis to a considerable degree (13); this work is reviewed at length in Section 7.2.3.

Elgin and Goodstein (6, 8) measured vapor pressures and heat capacities of ⁴He adsorbed on graphite, using in their studies at the California Institute of Technology a similar form of exfoliated graphite† as that employed by Bretz *et al.* (1–5) at the University of Washington. Elgin and Goodstein emphasized a region of lower coverage and higher temperatures, but also surveyed a considerable region of overlap with the University of Washington data; in that region the agreement between the two independent sets is completely within experimental error. Elgin employed both sets of measurements to generate extensive tables of the thermodynamic functions of ⁴He on graphite (13), which were used as basis for detailed interpretations of the various phases of the films. Of particular importance in this chapter is their analysis of the melting regime of He⁴ and of relevant theories of melting in two dimensions. These topics are discussed in Section 7.3.

A study of ³He monolayer heat capacities in the solid range was carried out by Hering and Vilches (4, 10) using calorimeters that had been used for previous studies of adsorption on graphite. Their results are quite similar to the ⁴He data in all respects, but perceptible quantitative differences are seen. The empirical Debye temperatures $\theta(2D)$ for ³He are appreciably higher than those of the heavier isotope at equal density n. There is a comparable but somewhat smaller isotope shift in the Debye temperatures of 3D solid He. The difference be-

† "Grafoil"; see Chapter 3, footnote on p. 53.

tween the isotope shifts in films and bulk is considered to be significant and is discussed in Section 7.2. The ^{3}He specific heats exhibit melting peaks at virtually the same densities and temperatures as those of ^{4}He, but the ^{3}He peaks are higher and narrower. The ^{3}He measurements were made to sufficiently low temperatures to permit sensitive estimates of the entropy by integrating the heat capacity curves. By drawing a smooth base line connecting the regions above and below the peaks, a rough estimate could be made of the entropy change on melting. The values ranged from $\Delta S/Nk \simeq 0.0093$ for $n = 0.078$ (Å)$^{-2}$ to $\Delta S/Nk \simeq 0.09$ for $n = 0.092$ (Å)$^{-2}$ monotonically increasing with film density. For ^{4}He at the lower density, the melting anomaly is too diffuse for a reliable baseline, but at $n = 0.093$ (Å)$^{-2}$ one obtains $\Delta S/Nk = 0.06$. These differences are roughly one-tenth of the melting entropies of the corresponding 3D solids. For example, in hcp ^{4}He at equivalent molecular density $v^{-2/3} = 0.095$ (Å)$^{-2}$, $\Delta S_m/Nk \simeq 0.62$; for bcc ^{3}He at $v^{-2/3} = 0.093$ (Å)$^{-2}$, $\Delta S_m/Nk \simeq 0.76$.†

The entropy changes associated with the film peaks are also small in comparison to the total entropies. In ^{3}He at $T = 4°$K, for example, S/Nk ranges from 0.61 to 0.38 as density increases from 0.078 to 0.092 (Å)$^{-2}$. Since it is virtually certain that the nuclear spins are disordered at the lowest temperatures of the experiment, these values do not include the ln 2 contribution of spin degeneracy.

Since the ^{3}He measurements were carried out to very low temperatures, it was possible to make a fairly stringent test of the 2D solid approximation for these films. Any appreciable effects of substrate structure in the form of partial registry would cause the heat capacity to fall below the T^2 law at low T (see Chapter 8). However, as can be seen in Fig. 7.4, the data of Hering and Vilches follow the quadratic curves within experimental scatter down to the minimum temperatures of the measurements, for a ratio $T/\theta(2D)$ as low as 3×10^{-3}.

7.1.2 Second Layer of ^{4}He on Graphite

Bretz (5) has obtained evidence for the existence of a 2D solid second layer of ^{4}He on graphite, at coverages near second layer com-

† The melting entropy changes were calculated from the Clausius–Clapeyron relation, using the data of Ref. 15.

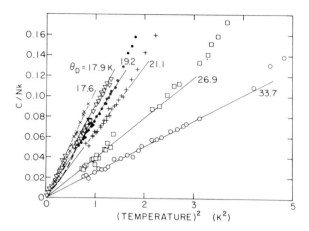

FIG. 7.4   The low temperature specific heat versus T^2 of high density ³He adsorbed on graphite (4, 10). The characteristic temperatures of the samples refer to the 2D Debye model. Monolayer densities n (Å^{-2}) are: (×) 0.078; (▽) 0.079; (●) 0.080; (+) 0.082; (□) 0.087; (○) 0.092; (– – –) 0.077.

pletion and temperatures below $T \simeq 3°K$. The second layer solid is quite distinct thermodynamically from the first layer, having an appreciably lower maximum density due to the much weaker substrate field. Strong peaks in the heat capacity at coverages of 2–3 layers were identified with melting of the second layer. These peaks trace out a melting line which parallels the melting line of the first layer, but is displaced to lower temperatures. A tentative explanation for the displacement is that the atoms in the partial third layer can facilitate melting by exchange with the atoms of the second layer. An analogous increase in mobility was seen by Rollefson in nmr of ³He just at completion of the first layer (16). Bretz has confirmed that the first layer retains solid character beyond the melting region of the second layer by observing that a strong melting peak occurs near 9°K, similar in form and temperature to the melting anomalies of single layer films.

7.1.3   Neon on Graphite

Neon monolayer heat capacities exhibit strong and very narrow peaks at a fixed temperature and over a range of intermediate coverages. These results, obtained by Huff and Dash (17, 18) and illustrated in Fig. 7.5, are strongly suggestive of a monolayer triple point in which 2D solid, liquid, and vapor phases (together with 3D vapor) are in equilibrium. The instrumental resolution of the individual heat capacity points was only a factor of two or three better than the width of the experimental curves, so that the observed peaks are consistent with ideal δ-function singularities degraded to some extent by instrumental

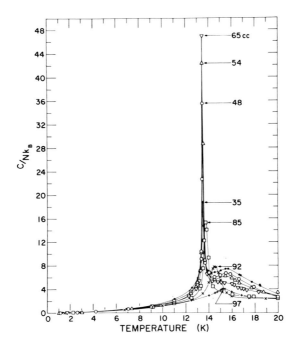

FIG. 7.5 Heat capacities of neon on exfoliated graphite (17, 18), indicating a triple point at intermediate coverage. Monolayer capacity 115 cc STP.

	N (cc STP)	x
●	35	0.303
○	48	0.416
△	54	0.471
▽	65	0.563
□	85	0.739
×	92	0.800
+	97	0.844

limits and perhaps some substrate heterogeneities. Previous measurements of heat capacities of Ne films adsorbed on graphitized carbon black, by Steele and Karl (19) and by Antoniou *et al.* (20) had disclosed strong broader anomalies in the first layer films, but only one sample coverage was measured in each of these studies.

The heat capacities at temperatures below the triple point are approximately Debye-like over a range of temperatures, but they fall below the T^2 law at low T. This indicates a gap in the density of states at low frequencies, which could be caused by partial registry with the substrate structure (see Chapter 8). Registry can impress the regularity of the substrate and force long-range order in the film, removing the divergence in positional long-range order that occurs in two-dimensional solids at finite T. As discussed at length in Section 7.3, according to former theories this could cause profound changes in the nature of melting in the film. These views are now in the process of revision, and it is not clear what role partial epitaxy will have in newer theories of melting. But its relevance nevertheless seems clear, especially as the nature of the heat capacity peaks seems unique in all monolayer systems studied up to the present.

At coverages above the intermediate range the "triple point" peaks progressively broaden and move to higher temperature. Contrast this to a system undergoing an ideal first-order transition in a container of fixed area or volume, which would exhibit a stepped heat capacity anomaly (Fig. 4.5) when the quantity in the cell is more than enough to fill it completely with liquid at the triple point. One possible explanation for the actual shapes of the Ne peaks is coverage variations throughout the sample due to incomplete sample annealing in preparation. Evidence for coverage variation is seen in the rounded anomalies at higher temperature, which are attributed to gas–liquid condensation. Discussion of effects of coverage variations in phase condensation is given in Chapter 9, Section 9.6.

7.1.4 N₂ Monolayers and Multilayers

Several years ago Morrison *et al.* (21) observed strong heat capacity anomalies in N_2 multilayers adsorbed on TiO_2 powder, and attributed them to some sort of melting phenomenon. Their results are shown in

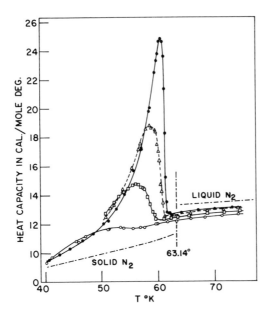

FIG. 7.6 Heat capacities of N_2 films adsorbed on TiO_2 (21). Coverages specified relative to the monolayer capacity V_m of a BET isotherm are: (●) 4.8; (△) 4.0; (□) 3.1; (○) 2.2.

Fig. 7.6. The relationship to the bulk melting point is obvious, and it appears that the peaks approach the bulk value and become sharper as thickness is increased. No anomalies were observed within the experimental range at coverages below two layers. These results are interesting in the light of more recent studies of N_2 adsorbed on graphite by neutron diffraction (22, 23). (See Fig. 7.7). At coverages just below one full monolayer a close packed structure is observed, with lattice spacing smaller than that of the basal plane graphite substrate. (A registered phase having larger lattice spacing was also observed: see Chapter 8.) This 2D solid phase has a sharp diffraction peak at low temperature, indicating long-range crystalline order. As the temperature rises above 60°K the peak broadens and moves to lower angles, indicating lateral expansion with decreasing range of order. The additional atoms forced out of the first layer by the 2D thermal expansion are evidently promoted into the second layer and

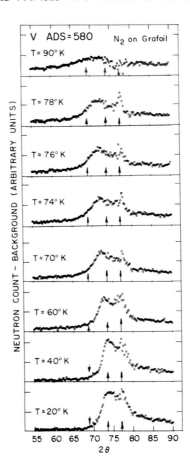

FIG. 7.7   The neutron diffraction pattern of N_2 adsorbed on Grafoil, with the basal planes oriented parallel to the scattering plane (23). At temperatures below $40°K$ and surface densities slightly less than full monolayer coverage, there is a sharp diffraction peak at scattering angle $2\theta = 73.6°$, which corresponds to the spacing of atoms in the (111) plane of solid N_2. As the temperature is raised the peak shifts to smaller angles and broadens, indicating thermal expansion and gradual loss of long range order. The other arrows in the figure mark the locations of the peaks due to the $\sqrt{3}$ epitaxial phase and the (002) line of misoriented crystallites.

the 3D vapor. The disappearance of the 2D solid is continuous within experimental resolution.

At substantial second layer coverages and low temperatures, the neutron diffraction patterns show crystalline order in both the first and second layers. The second layer is registered with the first, its atoms residing in the deep sites of threefold symmetry with respect to the first layer atoms. As the temperature is raised the second layer disorders, while the first retains solid character up to its own region of gradual melting.

7.1.5 Other Gases and Substrates

Multilayer "melting" anomalies have been seen in Ar (24), methane (25), and H_2O (26) films resembling the peaks observed by Morrison et al. in the N_2–TiO_2 systems (see preceding section). The Ar and methane films were adsorbed on TiO_2 and the H_2O films on porous glass; with minor variations all showed characteristics similar to N_2, i.e., strong asymmetric peaks centered at temperatures just below the bulk melting point. No anomalies were seen in films thinner than about three layers. The peaks tend to sharpen and shift to higher temperature as coverage is increased (but the peaks of the H_2O films remain at the same temperature). The anomalies are not understood at present.

Several monolayer solids may have been observed by vapor pressure isotherms. In their studies of Kr, Xe, and methane adsorbed on exfoliated graphite, Thomy and Duval observed several definite features during the formation of the first layer (27). These features were identified with a succession of gas, liquid, and solid phases, with vertical sections in the regions of two-phase coexistence. The upper vertical sections near full coverage (see Fig. 6.1) are reasonably associated with liquid–solid coexistence and the curved region above it to compression of the solid. In some of the films there is a second short vertical section attributed to "rearrangement" of the solid (28), tentatively identified as an epitaxial phase of $\sqrt{3} \times \sqrt{3}$ structure (see Chapter 8). No positive identifications have yet been made via LEED or other techniques, and it is not known whether either "solid" phase is an epitaxial array.

Up to the present, there have been no definite observations of non-epitaxial solid monolayers adsorbed on any substrates other than graphite. Several types of systems (29, 30) display vertical risers in vapor pressure isotherms similar to the midsections of Fig. 6.1 below $T_c(2D)$, but these are apparently signals of vapor–liquid and not liquid–solid coexistence. The reason for the absence of a second riser in the isotherms is not clear, although it is plausible that slightly poorer adsorption uniformity or film equilibration could blur such minor features of the overall curves.

7.2 THEORY OF SOLID MONOLAYERS

7.2.1 Thermodynamic Properties in the Harmonic Approximation

The simple theory of 2D solid monolayers can be constructed quite directly from the conventional 3D theory (31–33), with the addition of a term for the binding to the substrate and a change in the density of states. Although these modifications are straightforward, they may not be well known to all those interested in the field of adsorption, and therefore the development is carried out in some detail. The discussion in this section will also serve as an introduction to the next two divisions, which deal with much more recent developments.

Assume N adsorbed atoms in equilibrium on a smooth substrate of area A. The adatom–substrate attractive energy is taken to be ϵ_0 per adatom and the static adatom–adatom interaction energy is some function $u(a)$, where $a \equiv A/N$. The total Helmholtz free energy can be written as a sum of contributions due to the static terms and the vibrational excitations as

$$F = -N\epsilon_0 + Nu(a) + F_{\text{vib}}. \qquad (7.2.1)$$

The excitations are assumed to be harmonic vibrations of frequencies ω_i, where $i = 1, 2, \ldots, 2N$. Adapting the usual 3D theory, we find

$$F_{\text{vib}} = -kT \ln Z_{\text{vib}} = -kT \sum_i^{2N} \ln \left\{ \sum_{n_i=0}^{\infty} \exp\left[-\beta\hbar\omega_i(n_i + \tfrac{1}{2})\right] \right\}$$

$$= -kT \sum_i^{2N} \ln \left\{ \exp\left(-\beta\hbar\omega_i/2\right)/[1 - \exp\left(-\beta\hbar\omega_i\right)] \right\}$$

$$= (\hbar/2) \sum_i \omega_i + kT \sum_i \ln\left[1 - \exp\left(-\beta\hbar\omega_i\right)\right].$$

$$(7.2.2)$$

If the normal mode spectrum is quasicontinuous with density of states $g(\omega)$, the sums in Eq. (7.2.2) can be approximated as integrals and the free energy can be written in the form

$$F = -N\epsilon_0 + Nu(a) + (\hbar/2) \int_0^\infty \omega g(\omega)\, d\omega$$
$$+ kT \int_0^\infty g(\omega) \ln\,(1 - e^{-\beta\hbar\omega})\, d\omega. \qquad (7.2.3)$$

Equation (7.2.3), with the exception of the substrate binding energy, is formally identical to the equation for the free energy of a 3D harmonic solid. Differences between the 2D and 3D models must therefore depend upon their specific distribution functions.

In the simplest (Einstein) model, $g(\omega)$ is a δ-function. The only difference due to dimensionality is seen in the coefficient equal to the proper total number of modes. As discussed in Chapter 8, the Einstein model has some physical relevance to films in which the periodic potential plays an appreciable role. The model is inconsistent, however, with a smooth substrate approximation. Here it is more appropriate to use a "2D Debye" model, i.e., a planar analogue of the familiar 3D theory.

For the 2D Debye model we assume that $g(\omega)$ has the frequency dependence corresponding to 2D phase space, i.e., proportional to ω up to a cutoff frequency ω_D such that $\int_0^\infty g(\omega) = 2N$; thus we obtain

$$g(\omega) = 4N\omega/\omega_D^2, \qquad 0 \lesssim \omega \lesssim \omega_D,$$
$$= 0, \qquad \omega > \omega_D. \qquad (7.2.4)$$

Substituting in Eq. (7.2.3), we obtain for the vibrational part of F

$$F_{\text{vib}} = \tfrac{2}{3}Nk\omega_D + (4NkT/\omega_D^2) \int_0^{\omega_D} \omega\, d\omega \ln\,(1 - e^{-\beta\hbar\omega}). \quad (7.2.5)$$

Defining a characteristic temperature $\theta = \hbar\omega_D/k$ and changing variable to $y = \beta\hbar\omega$, we have

$$F_{\text{vib}} = \tfrac{2}{3}Nk\theta + 4NkT(T/\theta)^2 \int_0^{\theta/T} y\, dy \ln\,(1 - e^{-y}). \quad (7.2.6)$$

At low temperatures $T/\theta \ll 1$ the integral can be approximated by a simple analytic expression, leading to

$$F(\text{low } T) \cong -N\epsilon_0 + Nu + \tfrac{2}{3}Nk\theta - 4.808NkT(T/\theta)^2. \quad (7.2.7)$$

The corresponding entropy and heat capacity are

$$S = -(\partial F/\partial T)_{N,A} = 14.424Nk(T/\theta)^2, \qquad (7.2.8)$$

$$C = T(\partial S/\partial T)_{N,A} = 28.848Nk(T/\theta)^2. \qquad (7.2.9)$$

The chemical potential μ and the 2D pressure ϕ both involve changes in the areal density and therefore in the interaction energy u. If we also allow for a density dependence of θ (quasi-harmonic approximation), such variations can be described in terms of a "2D Gruneisen constant" γ:

$$\gamma \equiv -(\partial \ln \theta/\partial \ln A)_{T,N} = -(\partial \ln \theta/\partial \ln a)_T.$$

Treating F as an implicit function $F(T, N, A, \theta)$, we expand the usual thermodynamic derivatives to obtain

$$\mu = (\partial F/\partial N)_{T,A} = (\partial F/\partial N)_{T,A,\theta} + (\partial F/\partial\theta)_{T,A,N}(\partial\theta/\partial N)_{T,A}$$

$$= (\partial F/\partial N)_{T,A,\theta} + (\gamma\theta/N)(\partial F/\partial\theta)_{T,A,N}, \qquad (7.2.10)$$

$$\phi = -(\partial F/\partial A)_{T,N} = -(\partial F/\partial A)_{T,N,\theta} - (\partial F/\partial\theta)_{T,N,A}(\partial\theta/\partial A)_{T,N}$$

$$= -(\partial F/\partial A)_{T,N,\theta} + (\gamma\theta/A)(\partial F/\partial\theta)_{T,N,A}. \qquad (7.2.11)$$

Applying these relations to the low temperature approximation for F yields us

$$\mu = -\epsilon_0 + u - a\,du/da + \tfrac{2}{3}\gamma k\theta + 9.616\gamma k\theta(T/\theta)^3, \qquad (7.2.12)$$

$$\phi = -du/da + \tfrac{2}{3}(\gamma k\theta/a) + 9.616(\gamma k\theta/a)(T/\theta)^3. \qquad (7.2.13)$$

Equation (7.2.13) is the equation of state of the film. By equating μ in Eq. (7.2.12) to that of a 3D gas, we can obtain the vapor pressure of the film. If instead of using Eq. (7.2.12) directly we first combine it with Eq. (7.2.13) and then equate chemical potentials of film and vapor, we obtain an expression which is both compact and exposes a direct dependence on the equation of state of the monolayer:

$$P = (kT/\lambda^3) \exp\left[(\phi a - \epsilon_0 + u)/kT\right]. \qquad (7.2.14)$$

At temperatures $T \gtrsim 0.1\theta$, the integral in Eq. (7.2.6) must be evaluated numerically. The qualitative character of the 2D heat capacity at intermediate temperatures is similar to the familiar 3D curve, both having a general sigmoid shape with an inflection point at $T \sim \theta$. The asymptotic high temperature value of $C(2D)$ is $2Nk$ for both the Debye and Einstein models, and for all 2D harmonic solids, i.e., those having arbitrary spectra of harmonic modes with temperature-independent frequencies.

Improvements on the Debye approximation can be made in successive stages. First we can recognize that the density of states must reflect the fact that the velocities of transverse and longitudinal sound are invariably quite different. Retaining the assumption that the sound velocity is independent of frequency and proportional to the density of states in 2D phase space, we obtain for the two branches distinctive spectral densities

$$g_{l,t}(\omega) = A\omega/2\pi c_{l,t}^2, \qquad 0 \lesssim \omega \lesssim \omega_{l,t},$$
$$= 0, \qquad \omega > \omega_{l,t}, \qquad (7.2.15)$$

where $\omega_{l,t}$ are the individual cutoff frequencies. With the normalization condition specifying that the total number of modes in each branch is equal to N, the maximum frequencies and characteristic "Debye temperatures" $\theta_{l,t}$ of each branch can be related to the sound velocities

$$\omega_{l,t}^2 = 4\pi n c_{l,t}^2, \qquad k\theta_{l,t} = (4\pi n)^{1/2}\hbar c_{l,t}, \qquad (7.2.16)$$

where $n = N/A$. The overall density of states combining both branches is linear in ω at low frequency as in the original model but now has a sharp discontinuity at ω_t, and then continues linearly with a reduced slope until ω_l (34). The net effect on the heat capacity is to change the detailed dependence on temperature at intermediate temperatures, but to preserve the T^2 dependence at low T. In monolayers of interacting atoms the spectral density can be expected to have considerable structure, with one or more peaks in each branch (33, 35, 36). Such characteristics become less important at long wavelengths. As the wave length becomes $\gg \sqrt{a}$ the individualities due to lattice structure and ranges of interatomic forces are lost, and the density of states becomes that of harmonic modes in a uniform 2D continuum. Thus a real monolayer on a smooth substrate will have a heat capacity varying as T^2 at sufficiently low temperatures. "Sufficiently" depends on the actual spectral density function of this film; in the He–graphite systems discussed in Section 7.1, the empirical range is $T \gtrsim 0.07\theta$.

7.2.2   Quantum Solids in Two and Three Dimensions

The validity of the harmonic approximation rests upon the assumption that the amplitudes of atomic motions are small fractions of the interparticle spacing, so that a series expansion of the potential energy in powers of the displacements may be truncated after the quadratic terms. The truncation is justified at low temperatures for most solids composed of strongly bound atoms but is questionable even at $T = 0$ for van der Waals solids, particularly those of relatively light atoms (37–39), and similar qualifications can be made with respect to monolayers of such atoms. For these materials the combination of weak interatomic potentials and low masses combine to increase the atomic separations beyond the positions of minimum potential energy and cause the single-particle potential wells to become markedly flattened in their central regions. Crystals of this kind are called "quantum solids."

The fundamental model was first presented by London (40) with specific application to liquid ^{4}He. It was assumed that each particle in the liquid is enclosed in a rigid cage composed of its nearest neighbors and that the atoms are individually free to move about within each cage, up to hard-core contact. As a result of confinement each particle has a quantum-mechanical zero-point kinetic energy which acts as an internal pressure tending to spread the particles further apart, even at $T = 0$. The "zero-point pressure" is opposed by the interatomic attractive potential energy which, in the case of He, is not strong enough to cause solidification. The original idea of a quantum-mechanical kinetic energy due to confinement in an atomic neighbor cage is of recognized importance in low-mass van der Waals solids as well as in liquid He, but current theory is much more complicated than the original model. The complications are due primarily to correlation effects between neighboring atoms and to the detailed dependence of the atomic wavefunctions on the interatomic potential. The theory of quantum crystals is still under development; nevertheless the basic ideas are readily adaptable to monolayer systems. Some qualitative features can be seen by inspecting a simplified model first discussed by Cole (41) in a 3D context and later applied to He monolayers by Dash (42).

We assume a dense monolayer composed of N hard spheres of mass m. If each particle moves freely within a cage composed of its nearest neighbors, the ground state energy of the system is

$$E_0 = B[N(\pi\hbar)^2/2md^2], \qquad (7.2.17)$$

where B is a numerical factor due to correlations and d is the "rattle space" of each particle in its nearest neighbor cage. d can be expressed in terms of the fractional coverage x relative to a completed monolayer. If the effective atomic area is σ^2 the molar area of a film at completion is

$$A_0 = N_0\sigma^2/f^2, \qquad (7.2.18)$$

where f is a geometrical factor $\approx$ unity. By fractional coverage $x = A_0/A$ and by

$$d \cong (A/N_0)^{1/2} - \sigma, \qquad (7.2.19)$$

we obtain

$$E_0 = B[N_0(\pi\hbar)^2/2m][f^2x/\sigma^2(1 - f\sqrt{x})^2]. \qquad (7.2.20)$$

Cole showed that the high density approximation for the energy of the 3D hard sphere gas can be extrapolated to join smoothly with expressions for low density hard-sphere quantum gases (43–45). The low density 3D gas theories are expansions in powers of the parameter ρa^3, where ρ is the volume density N/V and a is the hard sphere diameter. Schick (46) showed that the analogous 2D expansion parameter is not na^2 but $(-\ln na^2)^{-1}$; Campbell and Schick (47) were then able to devise a smooth interpolation formula for the ground state energy of the hard disk system between the high density expression and the leading term of Schick's low density expansion.

The dependence of E_0 on film density gives rise to a zero-point spreading pressure ϕ_0;

$$\phi_0 \simeq -\left(\frac{\partial E_0}{\partial A}\right)_{T\sim 0} = \frac{B(\pi\hbar)^2}{2m} \frac{f^4x^2}{\sigma^4(1 - f\sqrt{x})^3}. \qquad (7.2.21)$$

This quantum mechanical spreading pressure provides a mechanism for the transmission of local density fluctuations along the surface. Fluctuations of relatively long wavelength can be analyzed into compressional waves, with phase velocity

$$c_i = (\partial\phi_0/\partial\rho)^{1/2}, \qquad (7.2.22)$$

where $\rho = Nm/A$ is the areal mass density. These density waves are collective modes of the entire system, and they have a density of states at small wavevectors which conforms to the usual phase space factor for the normal modes of 2D matter. Then we obtain via Eq. (7.2.22) for the phase space relation between sound velocity and characteristic temperature of the low-lying longitudinal modes, the expression

$$\theta_l = \frac{B^{1/2}\pi^{3/2}\hbar^2 x(4 - f\sqrt{\bar{x}})^{1/2}}{mk\sigma^2(1 - f\sqrt{\bar{x}})^2} . \tag{7.2.23}$$

There is no comparably simple theory for the transverse modes, but it is reasonable to assume that their phase velocity scales with the longitudinal velocity, at least in the high density range.

Other thermodynamic properties are readily derived. The vapor pressure for example, can be obtained from the chemical potential, which at low temperatures is just the derivative of the energy with respect to particle number;

$$\mu \simeq -\epsilon_0 + dE_0/dN, \tag{7.2.24}$$

where ϵ_0 is the single-particle substrate binding energy and E_0 is given by Eq. (7.2.20). Equating μ to μ_v of a 3D ideal gas, we obtain the vapor pressure equation

$$P = \frac{kT}{\lambda^3} \exp \left\{ -\frac{\epsilon_0}{kT} + \frac{B(\pi\hbar)^2}{2mkT} \left[\frac{(2\sqrt{a} - \sigma)}{(\sqrt{a} - \sigma)^2} \right] \right\}. \tag{7.2.25}$$

The hard-disk cell model is only a zeroth order approximation to real systems but serves to illustrate the importance of the quantum mechanical kinetic energy and how it contributes to thermodynamic properties. Kinetic energy effects were included in a much more realistic calculation by Campbell et al. (48), which was directed specifically to the properties of ^{4}He films adsorbed on graphite, at densities around completion of the first layer. They estimated the zero point kinetic energy due to He–He interactions according to the hard-disk interpolation formula of Campbell and Schick (47). An additional kinetic energy contribution due to motion normal to the substrate was included by the adoption of the single-particle ground state energies for ^{4}He and ^{3}He which had been previously obtained by Hagen *et al.* (49). The potential energy of He–He interactions was obtained by averaging the empirical 6–12 pair interactions with neighbors over single-particle cells in a triangular array. Numerical results were obtained for the ground state energy, chemical potential, monolayer completion density, and $\theta(2D)$, allowing the hard-core diameter to be varied to obtain the best fit to the film data. Agreement with measured monolayer completion densities of both isotopes was provided by the value 2.35 Å, in close agreement with the empirical helium gas potential parameter $\sigma = 2.56$ Å. Corresponding estimates for the 2D Debye temperatures are somewhat higher than the measured values, but seem within reasonable agreement, owing to uncertainties involved in estimating the velocity of transverse waves. A variational calculation by Novaco (50) gave strong support to the localized model and its neglect of exchange interactions, and yielded Debye temperatures in good agreement with experiment.

The ^{3}He and ^{4}He data show that the isotopic mass dependence of $\theta(2D)$ varies nearly as m^{-1} whereas $\theta(3D)$ varies approximately as $m^{-1/2}$, suggesting that quantum effects are more important in He monolayers than in solid He. In the hard-disk quantum solid model $\theta(2D) \propto m^{-1}$ [Eq. (7.2.23)], and in the equivalent hard-sphere approximation, one finds that $\theta(3D) \propto m^{-1}$ also. There are comparably elementary classical solid models which lead to the isotopic dependence $\theta \propto m^{-1/2}$. Instead of illustrating these we present a rather general treatment for the isotope shift in all solids, independent of their dimensionality, showing how the mass dependence varies according to their "quantum solid" character (51).

The theory depends upon the assumption that the characteristic temperature scales with the average low frequency sound velocity, both in 2D and 3D [see Eqs. (7.2.36) and (7.2.37)]. The sound velocities are related to the compressibilities; for the longitudinal modes,

$$c_{2l}^2 = -\left(\frac{A^2}{M}\right)\left(\frac{\partial \phi}{\partial A}\right)_{N,T}, \qquad c_{3l}^2 = -\left(\frac{V^2}{M}\right)\left(\frac{\partial P}{\partial V}\right)_{N,T}, \qquad (7.2.26)$$

where M is the total mass Nm. Both P and ϕ can be given as derivatives of the Helmholtz free energy, which at low T is essentially the energy itself, hence

$$
c_{2l}{}^2 = \frac{A^2}{Nm}\left(\frac{\partial^2 E}{\partial A^2}\right)_{N,T\to 0}, \qquad
c_{3l}{}^2 = \frac{V^2}{Nm}\left(\frac{\partial^2 E}{\partial V^2}\right)_{N,T\to 0}. \qquad (7.2.27)
$$

The mass dependence of the energy derivatives can be given for two extreme models, a harmonic crystal and a hard-sphere cell model of a quantum crystal. For a harmonic solid, the total energy can be written as the sum of a static structural energy and the vibrational zero-point energy as in Eq. (7.2.1). In the harmonic approximation the frequencies are independent of density, hence the energy derivatives in Eq. (7.2.27) involve only the static energy, which is independent of m. Thus, for both 3D and 2D in the harmonic approximation, $c_l \propto m^{-1/2}$. A similar argument can be given for the velocity of transverse waves, since c_t depends on a modulus of shearing elasticity which has a role equivalent to that of the inverse compressibility for c_l, and the same mass dependence obtains. Combining to form θ, we have

$$
\theta_{\mathrm{harm}}(2\mathrm{D}, 3\mathrm{D}) \propto m^{-1/2}. \qquad (7.2.28)
$$

The thermodynamic treatment is adaptable to the cell model of a hard-sphere quantum solid, as follows. In the low T limit E is the ground state energy of the many-body system, the lowest eigenvalue of the Schrödinger equation

$$
H\psi = E\psi, \qquad \text{where} \quad H = -\frac{\hbar^2}{2m}\sum_i^{3N}\nabla_i{}^2 + U(\mathbf{r}_1, \ldots, \mathbf{r}_N).
$$

We can formally express the sound velocities as

$$
c_{3l}{}^2 = \frac{V^2}{Nm}\left(-\frac{\hbar^2}{2m}\frac{\partial^2}{\partial V^2}\int \psi^*\sum_i \nabla_i{}^2\psi\, dV + \frac{\partial^2 U_0}{\partial V^2}\right), \qquad (7.2.29)
$$

where U_0 is the ground state potential energy. An equivalent expression obtains for $c_{2l}{}^2$, replacing V by A. Equation (7.2.29) can be applied directly to hard-sphere and hard-disk models without correlations. In this limit the energy is purely kinetic and each particle is confined to its own Wigner-Seitz cell. The total ψ is a product of ground state single-particle wavefunctions which are completely specified by the condition that they vanish at the cell boundaries. Therefore ψ is independent of m, leading immediately to the result that c_l^2 varies as m^{-2}. The argument carries through for transverse waves. Therefore, the mass dependence for the hard-core cell model is

$$\theta_{\text{hc,cell}}(2D, 3D) \propto m^{-1}. \tag{7.2.30}$$

This result, which contrasts with Eq. (7.2.28) is in a sense the limiting isotopic dependence of a "hard" quantum solid.

When the experimental results are inspected in the light of the two extreme models, it appears that He films come closer to the "hard quantum solid" limit than the 3D solids, which are more nearly harmonic [see Fig. 7.8]. The films and the solids increase toward the quantum solid limit as they are compressed. Both effects, of dimensionality and of increasing density, reflect changes in the near-neighbor dynamic

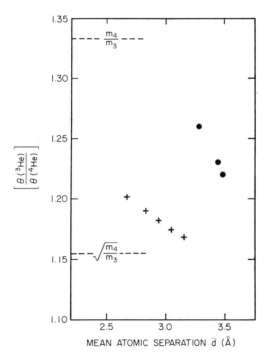

FIG. 7.8 Isotopic mass dependence of Debye temperature in hcp solid He (+) and monolayer solid He on Grafoil (●).

correlations. Equation (7.2.30) depends on two separate approximations, infinitely strong repulsions and the neglect of near-neighbor
dynamic correlations. The modern theory of 3D quantum solids, which
deals with both in more realistic fashion, yields effective single-particle
potentials which are essentially harmonic over an atomic displacement.
The success of the theory is evident from the near agreement of the
empirical isotope shift of 3D solid He with the harmonic ratio. But in
2D the dynamic correlations must be severely reduced by steric hind-
rance, motions of neighboring atoms to accommodate each other being
inhibited by the confinement to a plane (50).

7.2.3   Scaling between Films and Bulk Solids

The strong correspondence between the empirical characteristic
temperatures θ(2D) of monolayers and θ(3D) of hcp solid ^{4}He, as
illustrated in Fig. 7.1, is made on the basis of a simple $\frac{2}{3}$ power scaling
of the molecular volume. That scheme can provide only a semiquantitative basis for comparison, and the fact that the two sets of data lie on
virtually identical curves is probably accidental, for as we discuss below, there are several factors which enter in a more detailed comparison, and when these are included the data are no longer in coincidence.

An obvious consideration which is neglected in the $\frac{2}{3}$ power law is the
effect of crystal structure. If the comparison is based on a more detailed
microscopic basis, such as on a common scale of nearest-neighbor distances, then the two sets of data would not coincide, no matter what
structure the films might actually have, for that would introduce a
geometrical coefficient to the power law. For example if the monolayers
are actually of the triangular close packed (tcp) structure, as Elgin and
Goodstein (6, 8) have assumed, the proper scaling law adjusting the
hcp solid to the same interatomic spacing as the film is

$$a = 1.091v^{2/3}.$$

While there is no doubt that the crystal structure must play a role in
the films comparable in importance to that in the bulk, there is no
strong case at present for the assumption that the monolayer structure
is tcp. In bulk solid He there are both close-packed phases as well as
the less dense bcc occurring in different regions of the pressure–tem-

perature plane. The relative stability of the different phases is due to an extremely delicate interplay among several energy terms, and it has not yet been possible to calculate the terms with sufficient confidence to predict which phase is stable. The situation in the He films is probably just as delicate, together with the additional complication that the substrate texture may play a subtle yet determining role. It is to be hoped that the actual structure of the monolayers will be determined, but until it is known, the simple $\frac{2}{3}$ power law seems to be as reasonable as one based on any particular structure.

Apart from the foregoing issue, the similarity between $\theta(2D)$ and $\theta(3D)$ motivated Stewart et al. (9) to compare the elastic constants of the film and bulk. Their analysis and the more extensive developments carried out subsequently by Stewart (13) form the ground for the remaining part of this section.

Stewart et al. (9) showed that, by combining vapor pressure and heat capacity measurements in a general theory based on the Debye model, it would be possible to determine the principal elastic constants of the films and other mechanical properties, lending further insight to the similarities between the films and bulk solids. They first noted that the vapor pressure of a monolayer can be used to measure the "two-dimensional compressibility" K_2, defined in analogy with bulk matter as

$$K_2 \equiv -(1/a)(\partial a/\partial \phi)_T, \qquad (7.2.31)$$

where a is the molecular area and ϕ is the spreading pressure. Since the 2D chemical potential obeys

$$d\mu_{2D} = -s\,dT + a\,d\phi,$$

differentiation at constant temperature gives us

$$a = (\partial \mu_{2D}/\partial \phi)_T.$$

Substituting this expression for a in Eq. (7.2.31) yields us

$$K_2^{-1} = -(\partial \mu_{2D}/\partial a)_T. \qquad (7.2.32)$$

The chemical potential μ_f of a monolayer contains a relatively small term involving the volume v_f; if this is included the inverse compressibility is

$$K_2^{-1} = -(\partial \mu_f/\partial a) - v_f(\partial P/\partial a)_T, \qquad (7.2.33)$$

where P is the vapor pressure. The chemical potential of the film in equilibrium with its (ideal gas) vapor is given by P, T through

$$\mu_{\mathrm{f}} = \mu_{\mathrm{v}} = kT \ln (P\lambda^3/kT), \qquad (7.2.34)$$

where λ is the thermal de Broglie wavelength in the gas. Substituting (7.2.34) in (7.2.33) and neglecting v_{f} compared to v_{g} yields us

$$K_2^{-1} = -kT(\partial \ln P/\partial a)_T. \qquad (7.2.35)$$

In order to relate the elastic properties of films to their thermal properties, one needs two relations of the above type, since even in the simple Debye model there are two independent parameters, the transverse and longitudinal sound velocities. In both the usual 3D Debye model and in the 2D analogue, the heat capacity depends upon transverse and longitudinal velocities but with different weighting factors. Comparing 3D and 2D models, the characteristic temperatures in terms of the average velocities $\bar{C}_3$ and $\bar{C}_2$ are

$$\theta_3 = \frac{2\pi\hbar}{k}\left(\frac{3}{4\pi v}\right)^{1/3}\bar{C}_3, \qquad \frac{3}{\bar{C}_3} = \frac{2}{\bar{C}_{l3}} + \frac{1}{\bar{C}_{t3}}, \qquad (7.2.36)$$

$$\theta_2 = \frac{2\pi\hbar}{k}\left(\frac{1}{\pi a}\right)^{1/2}\bar{C}_2, \qquad \frac{2}{\bar{C}_2{}^2} = \frac{1}{\bar{C}_{l2}{}^2} + \frac{1}{\bar{C}_{t2}{}^2}. \qquad (7.2.37)$$

One of the necessary relations between elastic and thermal properties is Eq. (7.2.35), which was used together with the detailed tables of ⁴He monolayer thermodynamic functions that have been prepared by Elgin (14) from specific heat and vapor pressure data (3, 8). Since the second necessary variable was unknown over the full range of atomic separation, Stewart adopted the Cauchy conditions (31–33) as a substitute. The Cauchy conditions are fulfilled in systems in which every particle occupies a center of symmetry and all particles interact by purely central forces. In this event, Poisson's ratio $\sigma_3 = \frac{1}{4}$ is prescribed by elastic theory, and the sound velocities depend on a single elastic variable. This simplification had been applied to bulk solid helium data to estimate the volume compressibility from experimental values of the longitudinal sound velocity (52). But Stewart pointed out that the Poisson ratio $\sigma_3 = \frac{1}{4}$ corresponds to the Cauchy conditions together with zero external applied stress. In a medium under finite applied pressure, Poisson's ratio is a function of the compressibility and the pressure p, as

$$\sigma_3(pK_3) = \tfrac{1}{4}(1 + \tfrac{4}{3}pK_3)(1 - \tfrac{1}{3}pK_3)^{-1} \qquad (7.2.38)$$

The large compressibilities and high pressures required for solid helium make the correction to σ_3 important over the full experimental range of the solid. With the correct equation for the ratio in isotropic 3D media, the sound velocities are functions of the mass density ρ_3 as well as the pressure and compressibility,

$$C_{l3}{}^2 = \tfrac{9}{5}(\rho_3 K_3)^{-1}(1 - \tfrac{8}{9}pK_3), \qquad (7.2.39)$$

$$C_{t3}{}^2 = \tfrac{3}{5}(\rho_3 K_3)^{-1}(1 - 2pK_3). \qquad (7.2.40)$$

With these relations, Stewart found that bulk ^{4}He compressibility and calorimetric data could be brought into close quantitative agreement over the entire experimental range, whereas discrepancies of 20–25% exist if the zero stress Poisson's ratio is used instead. In 2D the formula analogous to (7.2.38) is

$$\sigma_2(\phi K_2) = \tfrac{1}{3}(1 + 2\phi K_2)(1 - \tfrac{2}{3}\phi K_2)^{-1} \qquad (7.2.41)$$

Application of the finite stress correction with Cauchy conditions to the film data produces an improvement in the comparison between the 2D Debye temperatures obtained calorimetrically and those computed from K_2. In contrast to the 3D situation, however, significant differences remain over the entire experimental range. Stewart concluded that the remaining discrepancies are probably due to a failure of the Cauchy conditions, which in turn may be caused by substrate structure effects. If the films are in partial registry with the substrate, some of the atoms are shifted from centers of symmetry. The proposal might be tested by a sensitive measurement of heat capacity at low temperature, searching for deviations from T^2 temperature dependence (see Chapter 8).

7.3 LONG-RANGE CRYSTALLINE ORDER

7.3.1 Mean Squared Displacements

One of the most intriguing aspects of the solid phases of monolayers involves the question of crystalline order and the regularity of their atomic structures. Theoretical interest in the problem of long-range crystalline order of lower-dimensional systems extends back many years. The original theory was presented by Peierls (53) as a qualitative argument based on a simple 1D model. The extension to 2D was outlined, and on this basis it was concluded that there cannot be long-range crystalline order in 1 or 2D atomic systems at any temperature above absolute zero. Landau reached the same conclusions, working from his general theory of second-order phase transitions (54, 55). In recent years there has been an upsurge of interest in the subject, with exact arguments for the absence of long-range order in lower dimensional systems under very general conditions (56–60). The conclusions are quite sweeping, relating not only to crystalline order but also magnetic, superfluid, and superconducting order. However, these conclusions have come under question, for it has been shown that the presence or absence of long-range order depends on the definition and physical significance of "order" (56, 61). The theoretical developments have a great deal of relevance to adsorbed films, and recent experiments have contributed to the revision of ideas. These changes have come at a fortunate time, during a period of increasing contact between theory and experiment, helping to stimulate the interplay. Here we outline the history and current status of theory as it relates to films on planar substrates: effects of substrate structure are postponed until the next chapter.

We begin with Peierls' 1D model, following a more detailed and quantitative treatment (62). Imagine a classical atomic chain of N atoms interacting by typical short-range forces. It can be assumed that the ground state at $T = 0$ is perfectly ordered, there being some uniform spacing d which is completely controlled by the interatomic forces and end conditions. As T is raised, the perfection of the order is destroyed; not only because the atoms move about their equilibrium positions in a random manner, but also because the equilibrium positions themselves become uncorrelated at large separations. This loss of "long-range crystalline order" happens because the equilibrium position of each atom is determined by its local environment, i.e., by the instantaneous positions of its neighbors rather than by any regular lattice fixed in space. For a quantitative gauge of the breakdown of crystalline order, one can compare the average separation between two atoms with the distance corresponding to the proper number of perfect lattice spacings. Thus, the instantaneous separation $x_n - x_0$ between the nth and 0th atoms fluctuates about nd at finite temperatures. The deviation $\delta_n \equiv (x_n - x_0) - nd$ can be calculated in terms of harmonic normal modes of the chain. As pointed out earlier, we note here that the harmonic theory does not require that the interatomic forces themselves be harmonic, but only that the chain be elastic on a scale involving many atoms. Then, within the limits of the harmonic approximation, the atomic displacement can be given as the sum (63)

$$x_n - nd = (Nm)^{-1/2} \sum_k V_k \exp[i(\omega_k t - knd)] \qquad (7.3.1)$$

where V_k is the amplitude of the mode of wavevector k and frequency ω_k. With the corresponding expansion for x_0 we obtain the deviation δ_n. The thermal average $\langle \delta_n{}^2 \rangle$ is found to be

$$\langle \delta_n{}^2 \rangle = (Nm)^{-1} \sum_k \langle |V_k|^2 \rangle \sin^2 (knd/2). \qquad (7.3.2)$$

The mean squared amplitudes of the normal modes are given, in the classical limit, by

$$\omega_k{}^2 \langle |V_k|^2 \rangle = c^2 k^2 \langle |V_k|^2 \rangle = k_{\mathrm{B}} T, \qquad (7.3.3)$$

where k_{B} is Boltzmann's constant. If the chain is very long the normal mode spectrum is quasi continuous and the discrete sum in Eq. (7.3.2) can be approximated as an integral over the spectrum. With density of states $g(k)$ in wavevector space and using the classical equation (7.3.3), we have

$$\langle \delta_n{}^2 \rangle = (4k_{\mathrm{B}}/Nm) \int_0^\infty \sin^2 (knd/2)[g(k)\, dk/c^2 k^2]. \qquad (7.3.4)$$

The density of states $g(k)$ depends upon the specific interatomic forces. We shall instead use an analytic expression which is asymptotically correct near $k = 0$ for all forces of realistically short range; i.e., the 1D analogue of the usual Debye approximation

$$
\begin{aligned}
g_{\mathrm{1D}}(k) &= L/2\pi, \qquad 0 \lesssim k \lesssim k_0, \\
&= 0, \qquad\qquad k > k_0,
\end{aligned}
\qquad (7.3.5)
$$

where $k_0 = 2\pi N/L = 2\pi/d$. Substituting in Eq. (7.3.4) and assuming c to be independent of k, we have

$$\langle \delta_n^2 \rangle = (2dk_{\rm B}T/\pi mc^2) \int_0^{k_0} \sin^2 (knd/2)(dk/k^2). \qquad (7.3.6)$$

Changing variable to $y = knd/2$, we find that Eq. (7.3.6) becomes

$$\langle \delta_n^2 \rangle = (nd^2 k_{\rm B}T/\pi mc^2) \int_0^{n\pi} \sin^2 y (dy/y^2).$$

The definite integral approaches the limiting value $\pi/2$ rapidly as $n \gtrsim 2$; with this limit the mean squared displacement is given by

$$\langle \delta_n^2 \rangle = nd^2 k_{\rm B}T/2mc^2. \qquad (7.3.7)$$

Thus, above $T = 0$, there is a monotonic loss of regularity in atomic positions with increasing separation, although the chain retains some degree of short-range order. For any finite T it becomes impossible to specify the positions of sufficiently distant atoms. There will be some n beyond which $\langle \delta_n^2 \rangle^{1/2}$ is greater than the interatomic spacing.

The methods of the 1D example are adapted to 2D systems as follows. The atomic positions become vector quantities $\mathbf{r}_n$ fluctuating about equilibrium locations $\mathbf{R}_n$, and the normal mode expansion takes the form

$$\mathbf{r}_n - \mathbf{R}_n = (Nm)^{-1/2} \sum_{\mathbf{k}} \mathbf{V}_k \exp [i(\omega_k t - \mathbf{k} \cdot \mathbf{R}_n)]. \qquad (7.3.8)$$

Now we take the projection of the relative deviation $\mathbf{r}_n - \mathbf{r}_0 - \mathbf{R}_n$ along some direction $\hat{e}$ lying in the plane of the array, and obtain for the mean squared deviation

$$\langle \delta_{n\hat{e}}^2 \rangle = (4/Nm) \sum_{\mathbf{k}} \langle |\mathbf{V}_k \cdot \hat{e}|^2 \rangle \sin^2 (\mathbf{k} \cdot \mathbf{R}_n/2). \qquad (7.3.9)$$

The features we wish to bring out can be seen by assuming a simple square lattice and choosing $\hat{e}$ along a principal lattice direction. Proceeding as before with the high temperature approximation and replacing the discrete sum by an integral over the 2D Debye density of states, we obtain

$$\langle \delta_{n\hat{e}}{}^2 \rangle = \frac{4Ak_BT}{(2\pi)^2 Nmc^2} \int_0^{k_0} \frac{dk}{k} \int_{-\pi}^{\pi} \sin^2 \left[\frac{knd \cos \theta}{2} \right] d\theta$$

$$= \frac{Ak_BT}{\pi Nmc^2} \int_0^{k_0} \frac{dk}{k} [1 - J_0(knd)], \qquad (7.3.10)$$

where J_0 is the Bessel function of the first kind (64). Integrating by parts, one obtains

$$\langle \delta_{n\hat{e}}{}^2 \rangle = (Ak_BT/\pi Nmc^2) \ln (\gamma k_0 nd/2) \qquad \text{as} \quad n \to \infty, \quad (7.3.11)$$

where $\gamma = 0.577 \ldots$ is Euler's constant (60). Thus for 2D as well as for 1D arrays there is no long-range crystalline order above $T = 0$. Although the divergence in the planar array is much weaker than in the chain, the mean squared displacement increases with n until a value is reached at which $\langle \delta_n{}^2 \rangle > d^2$; the order has a finite range. Because the dependence on n is only logarithmic, however, there can be extended short-range order, and if T is not too high the order can easily extend over the entire area of a large but finite array. But a qualitative difference exists between bulk matter and lower dimensional systems. If the same principles as used in the linear chain and planar array are applied to a three-dimensional model, it turns out that the mean squared deviation is independent of n. This means that there is no decay of correlation with distance, and as long as T is not excessive, crystalline order extends over an arbitrarily long range.

The preceding calculations involve a classical approximation for the normal mode amplitudes, suitable for the range of frequencies and temperatures such that

$$\hbar\omega_k = \hbar ck \ll k_B T. \qquad (7.3.12)$$

Outside of this range one must use the exact expression (65)

$$\langle |V_k|^2 \rangle = \frac{\hbar}{ck} \left[\frac{1}{e^{\beta\hbar ck} - 1} + \frac{1}{2} \right]. \qquad (7.3.13)$$

With Eq. (7.3.13) the expression for $\langle \delta_{n\hat{e}}{}^2 \rangle$ has no simple analytic solution, but numerical solutions show a logarithmic dependence on n similar to Eq. (7.3.12). This term is due to higher phonon occupations of the individual modes, while the constant term is due to zero-point motion. It is particularly interesting that the quantum zero-point motion, although it does contribute to the magnitude of $\langle \delta_n{}^2 \rangle$, is itself independent of n. Thus the destruction of long-range order in 1D and 2D systems at finite temperatures is a classical rather than a quantum phenomenon; we see that there can be long-range crystalline order in quantum crystals at $T = 0$, but as soon as T rises above zero, the thermal fluctuations destroy the long-range order.

For the 3D crystal the effect of zero-point motion is similar, adding a constant term to the expression for $\langle \delta^2 \rangle$. This change is of course less distinctive than in the 2D case, for in 3D there is no n dependence to the thermal part of the mean squared displacement.

Going beyond the Debye approximation, we understand that in realistic spectra there are significant peaks and singularities in the densities of states at moderate and high wavevectors. The effects of singularities on the mean squared displacement can be illustrated by treating them as sharp but finite width peaks superimposed on smooth distributions of Debye form. The contributions of the peaks and of the continuum are then additive, the continuum part being of the form calculated previously but now of smaller magnitude since the singularities contain a portion of the total number of modes of the array. The fraction can be denoted by some factor $\alpha < 1$. The contributions due to the singularities can be obtained directly by recognizing that since they have finite widths the oscillatory term $\sin^2 (knd/2)$ occurring in the integrals will be averaged, for at least moderate n, to the value $\frac{1}{2}$; hence the peaks yield a term independent of n. Therefore, the total mean squared displacement for a 2D array takes the form

$$\langle \delta_n{}^2 \rangle = \langle \delta^2 \rangle_{\text{sing}} + \alpha \langle \delta_n{}^2 \rangle_{\text{cont}}. \tag{7.3.14}$$

Equation (7.3.14) emphasizes the point that the decay of order with distance in 1D and 2D systems springs from the $g(k)$ peculiar to lower dimensional phase spaces. If singularities are present their effect is to *increase* the range of order by decreasing the fraction α. In the limit of the Einstein model, where the spectrum consists of only a δ-function singularity, there is no decay of correlation with distance for systems of any dimensionality. This is of course consistent with the implicit assumption of the Einstein model that particles are bound to sites of a fixed space lattice.

The primary relevance of the mean squared displacement to physical properties is to the diffraction of x-rays, neutrons, or low energy electrons from a film. One might at first expect that the absence of long-range order would imply that a 2D array could not have sharp coherent scattering peaks above $T = 0$, but would instead appear as an amorphous solid or liquid. But as pointed out earlier, the weak logarithmic divergence makes it possible for crystalline order to extend over an entire finite crystal or in what amounts to the same thing, over the full area illuminated by the radiation. This result was obtained from a detailed calculation by Imry and Gunther (64), who showed that Bragg-like peaks persist over a finite temperature range in a 2D harmonic model. Imry and Gunther also investigated the Mössbauer spectrum, finding a quasi-resonant line different in form from the usual Lorentzian resonant shape due to the large number of low energy phonons, but with an intensity comparable to that in a 3D lattice.

Questions of long-range order are also involved in other physical properties. However, it is important to note that the density fluctuations which destroy long-range order in the 2D solid do not introduce any singularities in the heat capacity. As Jancovici (65) has pointed out, the heat capacity and the fluctuations are both completely determined by the $g(k)$ of the system. In the Debye model, the specific heat has the asymptotic high temperature value $2k_B$ and yet $\langle \delta_n^2 \rangle_{n \to \infty}$ diverges at all finite temperatures.

7.3.2 Long-Range Directional Order and the Dislocation Theory of Melting

The primary impact of the theory outlined above would seem to concern the stability of the solid phase of monolayers (on quasi-smooth substrates). Since the familiar textbook idealization of a solid is a regular crystalline lattice of arbitrarily large extent, the absence of long-range order in 2D appears to predict that monolayers cannot have solid phases. But as pointed out by Gunther (66), the weak logarithmic dependence on size allows a *finite* sample to be ordered over a finite range of temperatures. Thus, the theory does not exclude the solid phase of laboratory-sized monolayer films, and is therefore in accord with experimental solid-like heat capacities; in this respect there is no qualitative difference between 2D and 3D matter.

However, qualitative differences dependent on dimensionality come to the fore when we try to apply the theory to melting behavior. The relevance of $\langle \delta_n{}^2 \rangle$ to the melting of 2D solids would seem to follow from the success of the Lindemann melting law of bulk matter (67). This empirical relation between the melting temperature T_m, Debye θ, atomic volume v and mass m,

$$\theta = \text{const } (T_m/mv^{2/3})^{1/2} \qquad (7.3.15)$$

can be derived by assuming that melting occurs when the rms displacement exceeds a small fraction of the interatomic spacing (63). With this interpretation the empirical fraction turns out to be nearly the same value, $\sim\frac{1}{8}$, for most simple solids. Since it is both reasonably constant and reasonable in magnitude it seems to provide a rationale for the fundamental correctness of this microscopic mechanism of melting. But it should be noted here that the correctness of this microscopic model is in no way established or generally accepted. There are several points of internal inconsistency in the so-called "derivation" of the Lindemann law and a number of important factors completely disregarded. At best, the model can only be taken as suggestive.† Furthermore, serious difficulties arise when the same ideas are used for a 2D melting theory. This is because in 2D the mean squared displacement is not independent of distance as it is in 3D, but rather increases monotonically with separation between atoms. Thus, if the 3D criterion is applied directly to 2D systems, it would imply that at intermediate temperatures the local environs of an atom are solid while distant regions are liquid; and that whether a particular region is solid or liquid would depend upon the location of the

† Among the criticisms that have been leveled at the conventional "derivation" of the Lindemann melting law are the following: (1) It assumes that the melting of the solid is independent of the disordered phase, whereas thermodynamic stability must depend upon the properties of both phases. (2) Interatomic forces are assumed to be harmonic, in which case breakdown would never occur. See Ubbelhode (68). We can now add a third of comparable seriousness: the logical difficulties that arise when the same arguments are applied to lower dimensional systems.

observer. With these observations Dash and Bretz proposed a modification which allowed a way around the paradox. They postulated that melting occurs on a mode-by-mode basis and is primarily associated with the breakdown of transverse modes. The criterion for breakdown was taken as a proportionality between the rms displacement $\langle \delta_n{}^2 \rangle^{1/2}$ and the wavelength nd of the particular mode. According to these assumptions melting in 2D is a continuous phase transition, beginning with the longest wavelengths at lowest temperatures and progressing to shorter wavelengths at higher T. When the modified theory is reapplied to 3D matter it properly predicts melting as a first-order phase transition; moreover, it recaptures the usual Lindemann law. The qualitative result of the theory, that melting in 2D is a continuous process, is consistent with the behavior of He monolayers on graphite. It was in fact those experimental results that led to the model, out of an attempt to connect the well-known prohibition of long-range order in 2D with the observation of broad heat capacity peaks apparently due to melting. However, in spite of this success, quite different arguments put forth at about the same time or somewhat later are more persuasive than the Dash–Bretz model, and although the question of melting in 2D is far from settled, it now seems to involve a rather different type of long-range order.

The alternative view was first suggested by Mermin (57), who noted that in spite of the divergence of the mean squared displacement, a 2D lattice retains long-range *directional* order over a finite range of temperatures. This type of order implies a correlation between the directions of the local lattice vectors at different locations in the crystal. If the equilibrium lattice sites are supposed to be $\mathbf{R} = n_1 \mathbf{d}_1 + n_2 \mathbf{d}_1$ and the actual instantaneous positions of the atoms are $\mathbf{r}(\mathbf{R})$, then directional long-range order can be gauged by the quantity

$$\langle [\mathbf{r}(\mathbf{R} + \mathbf{d}_1) - \mathbf{r}(\mathbf{R})] \cdot [\mathbf{r}(\mathbf{d}_1) - \mathbf{r}(0)] \rangle. \qquad (7.3.16)$$

If the particles were frozen to their equilibrium sites the value would be $d_1{}^2$. If there were no long-range directional order it would vanish for large $\mathbf{R}$. In fact the value calculated for a harmonic lattice turns out to be $d_1{}^2$ at all temperatures, even in the limit $R \rightarrow \infty$. This persistence of long-range directional order is in remarkable contrast to the destruction of long-range *positional* order at any temperature above absolute zero. Thus, in terms of the directional order the crystal structure is stable to arbitrarily high temperatures: it *never* melts. This surprising result is entirely consistent with the earlier finding that the heat capacity of the harmonic model remains solid-like at all T; both results follow from the assumption of perfectly harmonic forces.

The foregoing considerations indicate the need for an entirely different view of melting not only in 2D matter but also in bulk solids. New approaches to the 2D problem have been made by Kosterlitz and Thouless (63) and by Feynman (69). Their theories are essentially the same and are based upon the dislocation model of melting. In this model it is supposed that a liquid close to its freezing point has a local structure similar to that of a solid, but that in its equilibrium configurations there is some concentration of dislocations which can move under the influence of an arbitrarily small shear stress and so produce viscous flow. In the solid state there are no free dislocations in equilibrium, and so the system is rigid. Although isolated dislocations cannot occur at low temperatures in a large system (except near the boundary) since their energy increases logarithmically with the size of the system, pairs of dislocations with equal and opposite Burgers vector have finite energy and must occur because of thermal excitation. As the temperature is raised the mean length of a dislocation pair increases; melting occurs when the number of dislocation pairs of large separation becomes significant. The theory applied to 2D follows, according to the analysis of Feynman.

We consider the system at a low enough temperature for it to be in its solid phase, and consider the effect on the shear modulus of the thermal excitation of dislocation pairs.

The energy of a pair of dislocations of opposite Burgers vectors a distance r apart is

$$E_2 = 2L \ln r + \text{const}, \qquad (7.3.17)$$

where L depends on the elastic constants of the 2D medium. By contrast, the energy in the strain field of an unpaired dislocation is proportional to $\log A$, where A is the area of the medium, formally taken to be infinite. At low T, there are no unpaired dislocations to be found, but bound pairs, or dipoles, are thermally excited in equilibrium.

When a shear stress is applied, the dislocation pairs respond, both by polarizing and by stretching apart. These two effects, each proportional to r, contribute to decreasing the shear modulus (defined in the limit of zero applied stress). The material will therefore restore shear stresses only so long as $\langle r^2 \rangle$ is finite, where

$$\langle r^2 \rangle \sim \int r^2 \exp \left(-E_2/kT\right) r \, dr \sim \int r^{2(1-L/kT)} r \, dr. \quad (7.3.18)$$

We identify as T_M the temperature at which $\langle r^2 \rangle$ becomes infinite, which according to Eq. (7.3.18) is given by

$$T_M = L/2k. \quad (7.3.19)$$

Kosterlitz and Thouless give a formula for L in terms of the shear modulus and Poisson's ratio for the 2D solid, based on a standard result of 3D elastic theory. There is some ambiguity in the meaning of the 2D Poisson's ratio, however. Feynman has recomputed L from first principles in 2D, casting the result in terms of the longitudinal and transverse speeds of sound, c_l and c_t, as

$$L = (mb^2/4\pi a)[c_t^2 c_l^2/(c_t^2 + c_l^2)]. \quad (7.3.20)$$

where b is the Burgers vector, which can be approximated as $\sqrt{a}$. The two necessary elastic constants occur here in precisely the same combination as they do in the Debye temperature θ. Thus, only θ is needed to predict T_M at a given density, and one obtains

$$T_M = (2\pi)^{-2}(mk/8\hbar^2)a\theta^2. \quad (7.3.21)$$

This result has the same form as the Lindemann formula, with no adjustable constants. Furthermore, the theoretical values of T_M corresponding to the measured Debye temperatures of ^{4}He monolayers agree with the peak temperatures of the melting anomalies over the entire range of the measurements. The correspondence is somewhat less striking for ^{3}He films (4, 10), but nevertheless the trends are similar in their density variations.

The success of the Kosterlitz and Thouless–Feynman model points to the essential correctness of the dislocation mechanism and will perhaps stimulate renewed investigation of the theory of melting of 3D matter. However, it is important to keep in mind that there are several remaining puzzles in the melting of films. The sharpness of the transition seems to be beyond the present stage of the theory. Whether it is a first-order or higher-order phase change is expected to depend on interactions between the dislocations, since the elastic constants, which control their energies and hence their equilibrium concentrations, are themselves affected by finite concentrations of dislocations. Kosterlitz and Thouless estimated the effect by an iteration procedure amounting to a mean field approximation and found that the heat capacity has a narrow peak at T_M, finite in height and in all higher derivatives. Recognizing the type of error associated with mean field approximations in more familiar contexts, the authors suggested that a more accurate calculation would show a weak singularity in the specific heat at the transition.

The KTF theory is strictly two-dimensional, but there are strong indications that substrate structure plays an important role in the melting of real films. In all of the monolayer systems that have been observed to undergo some sort of "melting," the transitions appear to be continuous (or high order) in nonregistered films, and relatively abrupt (or first order) in registered films. (For the definition and discussion of "registry," see Chapter 8). In the high-density ^{3}He and ^{4}He phases, there is no evidence of registry, the T^2 heat capacity dependence extending down to the lowest temperatures of the experiments, as illustrated in Fig. 7.4. Registry would cause a changeover to a more rapid, exponential decrease at low T due to the absence of low-lying modes. Such exponential behavior is observed in Ne monolayers on graphite at low T, and these same films display strong melting peaks approximating δ-functions (17, 18). In Xe and N_2 monolayers on graphite there are well-defined triple points of $\sqrt{3} \times \sqrt{3}$ registered phases (see Chapter 8). In N_2 there is a high-density 2D solid phase which has a structure that is incommensurate with the graphite substrate, and this phase has a continuous melting transition, directly observed by the gradual broadening of the neutron diffraction lines, as shown in Fig. 7.7.

Notwithstanding the incomplete state of the theory, some aspects of the transition seem to be clearly understood. The T^2 heat capacities indicate rigidity at low temperatures, and the detailed comparisons between films and bulk show that the 2D solid model of helium films is essentially correct. The nmr line narrowing in ^{3}He monolayers on graphite indicates fluidity occurring at about the same temperatures and densities as those of the specific heat anomalies. Although the order of the transition and the inherent shapes of the peaks are not yet understood, the changes with density have been convincingly explained by Elgin and Goodstein (8). They found it significant that the peak shapes, which are relatively independent of density over much of the range, become impressively high within a narrow density interval near monolayer completion (see Fig. 7.2). Elgin and Goodstein attributed the strengthening of the peaks to "monolayer promotion," i.e., the excitation of some atoms from the first to the second layer due to the increase in spreading pressure upon melting, and supported their contention by a quantitative thermodynamic calculation. They assumed that the intrinsic shape of the peaks remains unchanged from the behavior at lower densities and calculated the additional contributions due to layer promotion and evaporation to the vapor phase by treating the transition as a diffuse first-order phase change, which allowed a calculation of the spreading pressure via a 2D Clausius–Clapeyron relation. The resulting compound peaks reproduce the experimental high-density shapes in every detail.

REFERENCES

1. M. Bretz, G. B. Huff, and J. G. Dash, *Phys. Rev. Lett.* **28,** 729 (1972).

2. R. J. Rollefson, *Phys. Rev. Lett.* **29,** 410 (1972).

3. M. Bretz, J. G. Dash, D. C. Hickernell, E. O. McLean, and O. E. Vilches, *Phys. Rev.* **A8,** 1589 (1973); **A9,** 2814 (1974).

4. S. V. Hering and O. E. Vilches, *in* "Monolayer and Submonolayer Helium Films" (J. G. Daunt and E. Lerner, eds.), p. 1. Plenum Press, New York, 1973.

5. M. Bretz, Ref. 4, p. 11.

6. R. L. Elgin and D. L. Goodstein, Ref. 4, p.˙35.

7. R. J. Rollefson, Ref. 4, p. 115.

8. R. L. Elgin and D. L. Goodstein, *Phys. Rev.* **A9,** 2657 (1974).

9. G. A. Stewart, S. Siegel, and D. L. Goodstein, *in Proc. Int. Conf. Low Temp. Phys., 13th* (R. H. Kropschot and K. D. Timmerhaus, eds.). Plenum Press, New York, 1974.

10. S. V. Hering, Ph. D. thesis, University of Washington, 1974.

11. G. Ahlers, *Phys. Rev.* **A2,** 1505 (1970).

12. D. F. Brewer and D. P. Grimmer, *in* "Monolayer and Submonolayer Helium Films" (J. G. Daunt and E. Lerner, eds.). Plenum Press, New York, 1973.

13. G. A. Stewart, *Phys. Rev.* **A10,** 671 (1974).

14. R. L. Elgin, Ph.D. thesis, California Inst. of Technol., Pasadena, 1973.

15. H. H. Sample and C. A. Swenson, *Phys. Rev.* **158,** 188 (1967); F. E. Simon and C. A. Swenson, *Nature (London)* **165,** 829 (1950); R. H. Sherman and F. J. Edeskuty, *Ann. Phys. (N.Y.)* **9,** 522 (1960).

16. R. J. Rollefson, private communication.

17. G. B. Huff and J. G. Dash, in *Proc. Int. Conf. Low Temp. Phys., 13th* (R. H. Kropschot and K. D. Timmerhaus, eds.). Plenum Press, New York, 1974.

18. G. B. Huff, Ph.D. Thesis, Univ. of Washington, 1973.

19. W. A. Steele and R. Karl, *J. Colloid Interface Sci.* **28,** 397 (1968).

20. A. A. Antoniou, P. H. Scaife, and J. M. Peacock, *J. Chem. Phys.* **54,** 5403 (1971).

21. J. A. Morrison, L. E. Drain, and J. S. Dugdale, *Can. J. Chem.* **30,** 890 (1952).

22. J. K. Kjems, L. Passell, H. Taub, and J. G. Dash, *Phys. Rev. Lett.* **32,** 724 (1974).

23. J. K. Kjems, L. Passell, H. Taub, J. G. Dash, and A. D. Novaco, *Phys. Rev.*

24. J. A. Morrison and L. E. Drain, *J. Chem. Phys.* **19,** 1063 (1951).

25. K. S. Dennis, E. L. Pace, and C. S. Baughman, *J. Amer. Chem. Soc.* **75,** 3269 (1953).

26. G. G. Litvan, *Can. J. Chem.* **44,** 2617 (1966).

27. A. Thomy and X. Duval, *J. Phys. Chim. Physicochim. Biol.* **67,** 1101 (1970).

28. A. Thomy, J. Regnier, J. Menaucourt, and X. Duval, *J. Crystal Growth* **13/14,** 159 (1972).

29. Y. Larher, *J. Colloid Interface Sci.* **37,** 836 (1971).

30. T. Takaishi and M. Mohri, *J. Chem. Soc., Faraday Trans. I* **68,** 1921 (1972).

31. M. Born and K. Huang, "Dynamical Theory of Crystal Lattices." Oxford Univ. Press (Clarendon), London and New York, 1954.

32. R. E. Peierls, "Quantum Theory of Solids," Chapter 2. Oxford Univ. Press (Clarendon), London and New York, 1955.

33. A. A. Maradudin, I. P. Ipatova, E. W. Montroll, and G. H. Weiss, "Theory of Lattice Dynamics in the Harmonic Approximation" *(Solid State Phys.,* Suppl. 3), 2nd ed. Academic Press, New York, 1971.

34. L. Brillouin, "Wave Propagation in Periodic Structures." Dover, New York, 1953.

35. E. W. Montroll, *J. Chem. Phys.* **15,** 575 (1947).

36. L. Van Hove, *Phys. Rev.* **89,** 1189 (1953).

37. C. Domb and J. S. Dugdale, *in* "Low Temperature Physics" (C. J. Gorter, ed.), Vol. II, Chapter XI. North-Holland Publ., Amsterdam, 1957.

38. L. H. Nosanow, *Phys. Rev.* **146**, 120 (1966).

39. B. H. Brandow, *Ann. Phys. (N.Y.)* **74**, 112 (1972).

40. F. London, *Proc. Roy. Soc. (London)* **A153**, 576 (1936); *J. Phys. Chem.* **43**, 49 (1939); F. London, "Superfluids," Vol. II, Chapter B. Dover, New York, 1964.

41. R. K. Cole, *Phys. Rev.* **155**, 114 (1967).

42. J. G. Dash, *J. Low Temp. Phys.* **1**, 173 (1969).

43. K. Huang and C. N. Yang, *Phys. Rev.* **104**, 767 (1957); **104**, 776 (1957).

44. T. T. Wu, *Phys. Rev.* **115**, 1390 (1959).

45. V. F. Efimov and M. Ya. Amus'ya, *Zh. Eksp. Teor. Fiz.* **47**, 581 (1964) [*English transl.: Sov. Phys.-JETP* **20**, 388 (1965)].

46. M. Schick, *Phys. Rev.* **A3**, 1067 (1971).

47. C. E. Campbell and M. Schick, *Phys. Rev.* **A3**, 691 (1971).

48. C. E. Campbell, F. J. Milford, A. D. Novaco, and M. Schick, *Phys. Rev.* **A6**, 1648 (1972).

49. D. E. Hagen, A. D. Novaco, and F. J. Milford, *in* "Adsorption-Desorption Phenomena" (F. Ricca, ed.), pp. 99–110. Academic Press, New York, 1972.

50. A. D. Novaco, *Phys. Rev.* **A7**, 678 (1973); **A8**, 3065 (1973).

51. J. G. Dash, *Phys. Rev. Lett.* **32**, 603 (1974).

52. J. H. Vignos and H. A. Fairbank, *Phys. Rev.* **147**, 186 (1966).

53. R. E. Peierls, *Ann. Inst. Henri Poincaré* **5**, 1771 (1935).

54. L. D. Landau, *Phys. Z. Sowjetun.* **11**, 26 (1937).

55. L. D. Landau and E. M. Lifshitz, "Statistical Physics," Chapter XV. Pergamon, Oxford, 1958.

56. H. E. Stanley and T. A. Kaplan, *Phys. Rev. Lett.* **17**, 913 (1966).

57. N. D. Mermin, *Phys. Rev.* **176**, 250 (1968).

58. P. C. Hohenberg, *Phys. Rev.* **158**, 383 (1967).

59. M. E. Fisher and D. Jasnow, *Phys. Rev.* **B3**, 907 (1971).

60. Y. Imry and L. Gunther, *Phys. Rev.* **B3**, 3939 (1971).

61. J. M. Kosterlitz and D. J. Thouless, *J. Phys. C: Solid State Phys.* **5**, 124 (1972); **6**, 1181 (1973).

62. J. G. Dash and M. Bretz, *J. Low Temp. Phys.* **9**, 291 (1972).

63. D. Pines, "Elementary Excitations in Solids," Benjamin, New York, 1963.

64. M. Schick, private communication. I am grateful to Dr. Schick for correcting the corresponding formula in Ref. 62.

65. B. Jancovici, *Phys. Rev. Lett.* **19**, 20 (1967).

66. L. Gunther, *Phys. Lett.* **25A,** 649 (1967).

67. F. A. Lindemann, *Z. Phys.* **11,** 609 (1910).

68. A. R. Ubbelhode, "Melting and Crystal Structure," Chapter 3. Oxford Univ. Press (Clarendon), London and New York, 1965.

69. R. P. Feynman, private communication to R. L. Elgin and D. L. Goodstein; reported in Refs. 6 and 8.

8. Epitaxial Monolayers

In epitaxial monolayers the adsorbed atoms are arranged in regular patterns in registry with the substrate structure.† The registry comes about because the adsorption energy varies along the surface, reflecting the atomic structure of the top-most layers of the substrate. The arrangements are not determined by the site positions alone. All of the basic film and surface parameters come into play: the depth, symmetry, and spacing of the sites; the adatom sizes and interaction

---

† There is no standard convention for the use of the term "epitaxy" or of several·others which have sometimes been put forth instead. According to Professor Harry C. Gatos, Editor of *Surface Science*, the term originally meant the growth of material B on substrate A, with B adopting the structure of A. Epitaxy has also been used for layers of A on substrate A, but present use generally refers to layers of a material grown with the same structure of the substrate whether or not they are the same material. Dr. Elizabeth Wood, whose article on the vocabulary of surface crystallography (1) has helped to standardize most of the common terms, states that "epitaxy" is evidently not universally accepted. "Registry is sometimes used to describe similar structures as, e.g., "perfect registry" (2); also "coherency" (3). Whether A and B are the same or different materials is occasionally described as "homo-epitaxy" or "hetero-epitaxy." Since there is no hard and fast convention both "epitaxy" and "registry" will be used in a general sense of a regularity of the monolayer arrangement in phase with the substrate structure; this usage is close to that of Lander and Morrison. Somewhat more detailed descriptions such as "partial registry" will be used here. When more explicit descriptions are required, such as the complexity of the match, we will resort to diagrams and crystallographic notation.

energies. These properties, together with the thermal energy and the coverage combine to establish whether or not a film is epitaxial and if so, with what structure and perfection. A few simple examples will illustrate these points. The most familiar and elementary epitaxial arrangement is of an adatom in every adsorption site, as in a Langmuir monolayer at full coverage. This registry comes about because in the Langmuir model the only equilibrium positions for the adatoms are on the sites, and there can be no more than one atom on each site. But with increasing realism more complicated arrangements ·become possible. Large adatom sizes may create a region of exclusion around each occupied site, and with blockage of adjacent locations the densest regular array becomes an alternation of occupied and unoccupied sites. Further complications arise with finite barrier heights between sites and attractive interactions between adatoms. Here one can imagine an infinite series of complex patterns, due to the interference between the substrate periodicity and the crystal structure of the adatoms alone.

Theoretical and experimental studies of epitaxy reach back many years but the field is far from mature. In this chapter the history of theoretical and experimental progress, the present state of the field, and some important problems for the future are outlined.

8.1   THEORY OF INTERACTING LATTICE GASES

The simplest lattice gas is the Langmuir model, which is solved completely by elementary statistical mechanics (Chapter 5). The next step in realism, the inclusion of interactions between adatoms on different sites, makes the theory much more difficult. The addition of intersite forces creates a large class of models, sometimes called "interacting lattice gases," which includes the well-known 2D Ising model as a special case. A small number of these have been solved exactly, while numerical and analytical methods have yielded approximate solutions of many others. Yet it must be kept in mind that, difficult as these models may be, they are still far from realistic in several respects. The further complications are discussed in subsequent sections.

The earliest formulation of an interacting lattice gas with its approximate solution was given by Fowler (4, 5). We state the assumptions of the model in the terminology of current usage. N localized classical atoms are adsorbed on a regular array of N_s adsorption sites. The adatoms have hard-core repulsions preventing multiple occupation of a site and also interact when occupying adjacent sites; the potential energy u between two atoms is taken to be

$$u = +\infty \qquad \text{if the atoms occupy the same site,}$$

$$u = -2\epsilon \qquad \text{if the two atoms are nearest neighbors,} \qquad (8.1.1)$$

$$u = 0 \qquad \text{otherwise.}$$

The principal computational problem consists of the evaluation of the configurational factor in the partition function, and this depends on the number of nearest-neighbor pairs. Fowler's calculation was based on the assumption of a random distribution, which is equivalent to a mean field approximation. Fowler's solution for $\epsilon > 0$ (attractive interactions) indicates that the monolayer splits into two phases at low temperatures, a dense phase with every site occupied and a low density random "gas" phase. The critical coverage for the transition is $x_c = (N/N_s)_c = \frac{1}{2}$ and the critical temperature $kT_c = -\epsilon/2c$, where c is the nearest neighbor site coordination number.

Peierls (6) adapted a method due to Bethe (7) for an improved calculation. In this method a representative atom and its nearest neighbors are treated exactly and all other atoms beyond the "first shell" are assumed to have a self-consistent average distribution. Peierls obtained

$$x_c = \tfrac{1}{2}, \qquad kT_c = \epsilon/\ln\left[c/(c-2)\right]. \qquad (8.1.2)$$

For a square array $c = 4$ and $kT_c = (1.442 \ldots)\epsilon$.

The mean field and Bethe–Peierls approximations continue to be useful. These methods together with series approximations and numerical techniques are currently applied to more realistic models of physical systems: some of these applications are discussed at the end of this section. However, at this point we turn to a discussion of the exact solutions of the Ising model and their applications to localized adsorption.

The relevance of the Ising model of magnetic interactions to a discussion of adsorption derives from its mathematical correspondence to a lattice gas of interacting atoms. Its importance is due to the fact that a great deal is known about Ising magnetic systems: in addition to the exact solution of 2D arrays, the properties of many other more complicated Ising-type systems have been studied. Extensive reviews (8–10) have presented the theory primarily in the context of magnetism. Here we outline the correspondence with lattice gases and present the theoretical results in terms of physical adsorption.

The Ising model of ferromagnetism assumes a regular lattice of particles with scalar spin coordinates $\sigma = \pm 1$ according to whether the spin points along some preassigned direction $(+1, \uparrow)$ or in the opposite direction $(-, \downarrow)$. If the interaction range is assumed limited to nearest neighbors, the pair energy can take one of two values according to whether the spins are parallel or antiparallel. The correspondence with a lattice gas is obtained by associating spin directions with site occupation numbers; by convention, $\downarrow$ corresponds to "occupied" and $\uparrow$ to "empty." The basic similarity between the two types of systems was observed long ago; however the detailed and complete transcription was made more recently by Lee and Yang (11).

Exact solutions exist for several 2D ferromagnetic and antiferromagnetic lattices in zero magnetic field. The first complete solution was given by Onsager (12), who studied rectangular arrays, i.e., square lattices with arbitrary coupling constants in the two directions. He showed that for finite coupling constants there is a second-order phase transition with a specific heat diverging logarithmically at the critical temperature. Techniques developed for rectangular arrays were successfully applied to other 2D lattices, and exact solutions are now known for triangular, honeycomb, and Kagomé (woven bamboo) lattices (8–10, 13, 14). With few exceptions all of these systems have transitions similar to the rectangular lattice. The exceptions are certain antiferromagnetic arrays for which no ordered state exists, an example being a triangular lattice with all three coupling constants equal. All of the known exact solutions are for zero magnetic field; translated into the language of the lattice gas model of a monolayer, zero magnetic field corresponds to a relative coverage $N/N_s = \frac{1}{2}$. In terms of the number of sites N_s the specific heat in the critical region can be written

$$C/N_s k = -A \ln |(T - T_c)/T_c| + B. \qquad (8.1.3)$$

Table 8.1 lists theoretical values for the critical temperatures kT_c/ϵ, the entropies S_c at the critical point, and the coefficients A of several plane lattices. It should be noticed that the values of S_c and A are quite similar for all systems undergoing phase transitions.

There are several additional properties of Ising systems of particular relevance to adsorption studies. The general shape of the specific heat is asymmetric away from the critical region. At low temperatures the temperature dependence is exponential (9) as exp $(-\text{const}/kT)$ while at $T > T_c$ the decrease is more gradual, and this is noticeable in Fig. 8.1. For finite systems the specific heat has a finite maximum value: the peak is rounded and shifted to slightly lower temperatures. For symmetric square lattices of $(m \times n) = N_s$ spins, with $m \to \infty$, n finite, the peak height is estimated to be

$$C_{max}/N_s k \simeq 0.4945 \ln (n) + 0.1879, \qquad (8.1.4)$$

i.e., with a coefficient of the logarithmic term equal to the value of A of the specific heat (12). Also, the peak location is shifted by an amount $O(\ln (n)/n^2)$. Since numerical calculations (15) for both m and n finite yield results which approach Onsager's expression for large N, we may estimate the peak height for an $m \times n$ finite monolayer on a square lattice by setting $n \simeq (N_s)^{1/2}$, obtaining for large

Table 8.1 Critical Parameters of Lattice Gases

Lattice	$kT_c/|\epsilon|$	$S_c/N_s k$	A	x_c
Square	1.1346	0.3065	0.4945	$\frac{1}{2}$
Triangular				
$\epsilon > 0$	1.8205	0.3303	0.4991	$\frac{1}{2}$
$\epsilon < 0$	0	—	—	—
Honeycomb	0.7953	0.2647	0.4781	$\frac{1}{2}$
Kagomé				
$\epsilon > 0$	1.0717	0.2805	—	$\frac{1}{2}$
$\epsilon < 0$	0	—	—	—

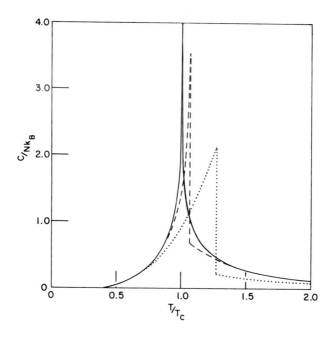

FIG. 8.1   Specific heats of the 2D Ising model-lattice gas in various approximations, as shown by Domb (9). ($\cdots$) Bethe–Peierls; ($----$) Kramers and
Wannier; (———) Onsager (exact).

systems

$$C_{\max}/N_s k \simeq \tfrac{1}{2} A \ln (N_s). \qquad (8.1.5)$$

Although there are no exact solutions for the Ising model in a finite
magnetic field, the general effect of an arbitrary field is known and
various series approximations have been developed (9, 16). In terms
of magnetic systems the presence of a field on a ferromagnetic array
causes a finite magnetization at temperatures above the zero field T_c;
for lattice gases this corresponds to coverages other than $x = \tfrac{1}{2}$. The
heat capacity singularity is destroyed, leaving a rounded peak. The
second-order phase transition in an antiferromagnet is preserved in a
finite field (17). It is predicted that the peak temperature is lowered in
the presence of a field (18); in lattice gas terms, the variation is

$$T_c(x_c) - T_c(x) = \text{const } (x - x_c)^2, \qquad \text{const} > 0. \qquad (8.1.6)$$

Additional rounding is expected if there are impurities in the system, e.g., variations in the coupling constants between the particles (19). If the variation has a narrow distribution width, of order $(N)^{-1}$, the specific heat becomes appreciably affected in the range $T - T_c = O(N^{-2})$. In addition the location of the maximum is shifted by an amount $O(N^{-1})$.

Considerable attention has been paid to more complicated Ising-type systems important to adsorption. As examples we mention numerical studies (20, 21) of hard-disk and hard-square gases with infinite nearest-neighbor repulsions, i.e., in which adjacent sites are blocked. These systems appear to undergo continuous transitions to dense ordered states which are similar to the "liquid–solid" transitions obtained for hard-sphere continuum fluids (22, 23). Series expansions and numerical methods have been applied to several cases in which there are finite first- and second-nearest neighbor interactions. A recent study (24) has extended the exploration by numerical methods to still more complicated systems; hard-core square and triangular lattice gases with weak long-range attractions and with blocking up to fourth neighbors. A variety of first- and second-order phase changes are found to occur, and within some ranges of the parameters, the resulting phase diagrams are similar to real 3D systems. Also of special importance are two recent studies of a classical (25) and a quantum-mechanical (26) 2D lattice gas with finite nearest-neighbor repulsions and second nearest-neighbor attractions, treated by the Bethe–Peierls approximation. The theoretical systems were modeled to imitate He monolayers on graphite, and the results of the calculations are discussed, together with the experiments, later in this chapter.

8.2 EXPERIMENTS ON EPITAXIAL MONOLAYERS

8.2.1 Direct Observations

The primary technique for direct observation of registered phases is low energy electron diffraction. The LEED technique has been more useful for chemisorbed films since the electron beams tend to desorb weakly bound layers (27).† However, there have been a number of successful studies of molecular gases on a variety of substrates exhibiting a wide range of phenomena including some of those treated in theoretical models.

A most extensive LEED study of strongly adsorbed molecular monolayers was done by Lander and Morrison (2) using the basal plane of a graphite single crystal as substrate. They observed many examples of epitaxial and nonepitaxial phases among the nineteen adsorbates investigated, and with some adsorbates they observed a sequence of phase changes as coverage and density were varied. A summary of substances investigated and phases observed is given in Table 8.2. For most of the films Lander and Morrison were able to deduce atomic arrangements of the atoms relative to each other and to the graphite lattice: in each case the monolayer structures were tested for consistency with the shapes and sizes of the molecules. Their observations of Br_2 are particularly interesting, for it exhibited four distinct 2D phases: lattice gas, liquid, and two different crystalline arrays. The lattice gas phase, seen at low coverage and $T \lesssim 250°K$, corresponds to semilocalized adsorption (i.e., with adatoms spending most of their time at well-defined adsorption sites) with an area of exclusion around each adatom. The exclusion was attributed to dipole–dipole repulsions arising from a static polarization moment of the bromine molecule. At lower temperature the diffraction pattern changed abruptly, indicating a denser and more amorphous structure. The transition from the 2D gas to liquid was deemed to imply a change in the state of ionization, and the authors cited as supporting evidence the decreasing ionization with increasing concentration seen in bromine–graphite intercalation compounds (28, 29). Further cooling caused an abrupt change in diffraction to a spot pattern of half-order spots, corresponding to a 2×2 crystalline array.‡ The molecular

† Recent technical developments increasing the sensitivity of beam detection promise to reduce necessary beam currents to much lower levels, permitting LEED studies on weakly bound gases.

‡ The conventional scheme for describing epitaxial structures is by relating the surface mesh of the adsorbed film to that of the substrate, by using the unit lattice vectors of the substrate as a basis. Thus, a monolayer with the same spacing and symmetry as the substrate has a 1×1 structure, and one with twice that spacing in both directions is referred to as 2×2.

Table 8.2   Two-Dimensional Crystal Phases of Monolayers on Graphite[a]

Material[b]	Pattern	Lateral bonding[c]	Structure
$FeCl_3$	Ring	Weak	Close packed (not bulk)
ZnI_2	Ring	Stable	Unknown (not bulk)
GeI_4	Spot	Weak	Close packed, in registry (not bulk)
GeI_2	Ring → spot	Stable	CdI_2-type close packed, in registry (bulk)
$C_6H_3Br_3$	Spot	Weak	Close packed, in registry (not bulk)
As_2O_3	Ring, spot	Stable	Graphite-type, not in registry (bulk)
Br_2 (3)	Spot	Weak	2×2 (?), in registry (not bulk)
(4)	—	Weak	4×4, in registry (not bulk)
Cs	Spot	Weak	2×2, close packed, in registry (not bulk)
Xe	Spot	Weak	($\sqrt{3}$ 30°)-type, close packed, in registry (bulk)

[a] LEED observations of Lander and Morrison (2).

[b] Two-dimensional crystal phases of the following were not obtained: hydroquinone, glycerin, naphthalene, benzene, P, AlI_3, CCl_4, $CHCl_3$, CBr_4, and $CHBr_3$.

[c] All materials with weak lateral bonding also gave a fuzzy ring "liquid"-like phase.

arrangement was not completely determined, but arguments favored an array with the molecular axis tilted out of the plane. A second crystal phase appeared at lower temperatures; this phase, which has lower 2D density, appeared to have a planar stacking structure with the long molecular axes parallel to the surface and the Br atoms of each molecule centered over adjacent adsorption sites.

Several of the Br_2–graphite phases observed by Lander and Morrison have similar properties to CO–(100)Pd. Tracy and Palmberg (30) studied this system by LEED, work function and surface Auger measurements. They observed four phases:

(1)   At low coverages (<0.4 monolayer), a random lattice gas in which each CO is localized on a site.

(2)   Just below $x = 0.5$, there is a liquid-like arrangement with short-range order.

(3)   At exactly $x = 0.5$ the monolayer has a centered 2×2 struc-

ture oriented at 45° with respect to the surface mesh, and in registry with the substrate.

(4)   At $x > 0.5$ the 2×2 structure is uniaxially compressed, forming a misregistered but ordered overlayer.

The binding energy, determined by both work function measurements and the isosteric heat of adsorption calculated from the temperature dependence of the diffraction pattern intensity showed a smooth linear decrease with increasing coverage at $x < 0.5$, a sudden decrease near $x = 0.5$, and an accelerating decrease at higher coverage.

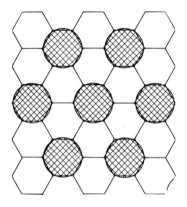

FIG. 8.2   The $\sqrt{3} \times \sqrt{3}$ structure of epitaxial Xe, N_2, and He monolayers on basal planes of graphite.

Some of the properties of Br_2–graphite and CO–(100)Pd may be complicated by changes in adatom–substrate chemisorption forces. Physisorption is less likely to involve marked changes in the binding; however, Xe monolayers have exhibited some of the same features seen in the bromine and carbon monoxide films. Lander and Morrison (2) observed an ordered Xe–graphite phase at $T \sim 90°$ and $P \sim 10^{-3}$ Torr. The structure was deduced as $\sqrt{3}$ R 30°, i.e., having a surface spacing of $\sqrt{3}$ larger than the graphite mesh and rotated 30° with respect to the graphite.† This structure is in registry with the substrate, but only $\frac{1}{3}$ of the absorption sites are occupied; it is the structure illustrated in Fig. 8.2. Heating the Xe a few degrees caused the

† This structure has several alternative designations in the literature, including "$\sqrt{3} \times \sqrt{3}$," "$\sqrt{3}$," and "$x_g = \frac{1}{3}$."

diffraction to change from a spot to a diffuse ring pattern, which was interpreted as a liquid-like phase with a roughly close-packed nearest-neighbor distribution. With further increase of temperature the rings disappeared; in this region the film appeared to be a 2D gas. The 2D crystal → 2D liquid and 2D liquid → 2D gas transitions were not abrupt; Lander and Morrison judged that they are probably second-order phase changes. The apparent second order nature of the transition may have been a nonequilibrium or instrumental effect. A recent reexamination of the Xe–graphite system by Suzanne et $al.$ (31) gave clear indication of a region of coexistence between the $\sqrt{3}$ registered phase and a 2D gas. In this study Auger electron spectroscopy was combined with LEED for quantitative measurements of coverage simultaneous with structural information, and this yielded conclusive evidence of two-phase equilibrium (see Fig. 8.3).

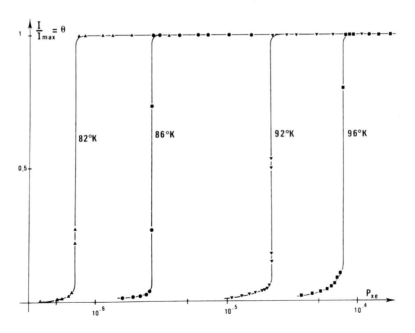

FIG. 8.3   Adsorption isotherms of Xe on graphite: Auger electron intensity versus pressure, measured by Suzanne et $al.$ (31). The step is well defined at $T \lesssim 99°$K, indicating a first-order transition between a low density 2D gas and a 2D solid. LEED examination shows that the solid has an epitaxial $\sqrt{3} \times \sqrt{3}$ structure.

Palmberg (32) carried out LEED, vapor pressure, and work function measurements on Xe–(100)Pd films at temperatures down to 77°K. Long-range ordering appeared only near monolayer coverage. The ordered phase was close-packed hexagonal, with nearest-neighbor spacing 4.48 Å, which is about 3% greater than bulk Xe. The Xe was not in registry with the smaller Pd mesh, but was oriented along the principal directions of the surface lattice. At lower coverage the Xe diffraction was diffuse, indicating a highly disordered structure. There was no evidence for a condensed liquid phase. The vapor pressure measurements yielded a linear decrease of q_{st} with increasing coverage. Both sets of data indicate that the Xe–Xe interactions are repulsive. The work function measurements were interpreted as indicating an appreciable static dipole moment associated with each Xe, with orientation normal to the surface, and Xe polarizability about twice that of an isolated atom. Dipole–dipole repulsion could account for the absence of a 2D liquid phase, while the ordering at high coverage could result from short-range repulsive interactions. The periodic potential of the substrate appeared to be relatively unimportant at the temperatures investigated, except for the tendency to orient the 2D crystal.

Tracy continued the study of CO, but with a change of substrates to (100)Ni and (100)Cu, combining LEED, Auger, and work function measurements (33, 34). He observed the complex changes with all of these techniques as the system underwent phase changes with variations of temperature and coverage. A particularly interesting observation was the coexistence of two distinct registered phases (34), a 2×2 simple square and a hexagonal phase, at temperatures in the range 300–400°K. Thus the strong binding ($q_{st} \simeq 1.3$ eV) of these films, which places them well within the range of chemisorption, does not necessarily prevent the achievement of rapid surface equilibrium at quite moderate temperatures and thermodynamic behavior that is normally thought of only in the context of physisorption.

A very recent study of N_2 monolayers on graphite (Grafoil) has been made by Kjems et al. (35) using neutron diffraction, the first application of this technique that has succeeded in identifying monolayer structures. A registered $\sqrt{3}$ phase was observed (see Fig. 8.4) over a fairly wide range of coverages, at temperatures below $\sim$85°K. As in

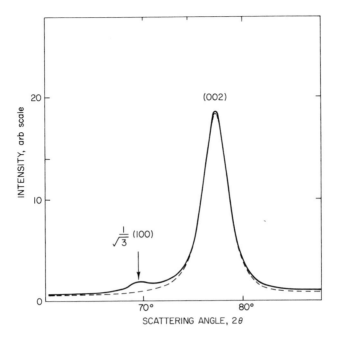

FIG. 8.4 Neutron diffraction indications of epitaxial $\sqrt{3} \times \sqrt{3}$ structure in N_2 monolayers on Grafoil (35). (———) Empty cell, showing (002) peak due to misaligned graphite crystallites (the majority of the crytallites are aligned with basal planes approximately parallel to the scattering plane); (64.5 Torr) (– – – –) cell with adsorbed N_2 near critical coverage, showing new diffraction peak corresponding to $\sqrt{3} \times \sqrt{3}$ epitaxial structure. $T = 78°K$.

the LEED study of Xe on graphite by Suzanne *et al.* (31), the neutron results suggest that the $\sqrt{3}$ registered phase–2D gas is a first-order transition at least over an appreciable range of T. A 2D solid phase was also observed, a dense hexagonal structure having nearly the same spacing as (111) planes of solid N_2.

Still other direct techniques are available for the study of registered phases. Mason and Williams (36) used low energy molecular scattering of He atoms to examine monolayers of H_2O, ethanol, and several other light organic molecules adsorbed on (001)LiF. For temperatures around 150°K, ethanol (C_2H_5OH) formed an ordered array, tentatively identified as oblique 2×2. At lower temperatures no peaks were observed, perhaps due to the formation of a partially disordered denser phase.

8.2.2 Indirect Observations

In the preceding section we discussed experiments employing techniques for the direct observation of ordered epitaxial arrays. While these probes yield unequivocal evidence of regular structures, they cannot produce complete descriptions of the film properties but must be supplemented by other techniques, such as calorimetry and nuclear magnetic resonance. As an illustration we survey in some detail the recent studies of He monolayers adsorbed on graphite at low temperatures, in which heat capacity and nmr were used to explore thermodynamic regimes identified with epitaxial phases (37, 38).

The identifications are based upon certain correspondences between the experimental heat capacities and theoretical models. The gross features of the ordered regimes are the emergence of strong and sharp heat capacity peaks in both ^{3}He and ^{4}He monolayers on Grafoil, the peaks being seen only within rather narrow ranges of temperature and coverage. The ^{4}He heat capacity at "critical coverage" is shown in Fig. 8.5; results with ^{3}He are nearly identical. It is most unusual to find the two He isotopes behaving in so similar a manner at low temperatures, for their behavior at low T is usually strongly influenced by quantum statistical effects. This similarity in the region of the peaks is one of the strong clues for the nature of the transition; it is clear that it must be dominated by just a few parameters; the adatom–adatom and adatom–substrate interactions, the number of atoms, and the temperature. Indeed, these factors are essentially just those of the Ising model. The next clue is the shape of the heat capacity peak, strongly resembling the exact Onsager curve (Fig. 8.1). Indeed, the behavior in the region of the peak is logarithmic, of the form of the theoretical equation (8.1.3), with nearly the same coefficient A on both high and low temperature sides of the peaks. The third clue is that critical coverage corresponds to the density of a $\sqrt{3}$ ordered array; this happens to be the densest regular structure consistent with the "size" of the He atom.

These results have raised a number of intriguing questions. It is

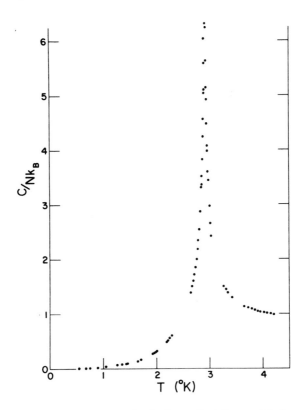

FIG. 8.5 Specific heat of ⁴He adsorbed on Grafoil at critical coverage for $\sqrt{3} \times \sqrt{3}$ epitaxial structure.

surprising that the shapes of the experimental peaks are so similar to theory, in view of the simplicity of the 2D Ising model. The models that have exact solutions so far do not include the $\sqrt{3}$ structure, or any of the following complications: forces beyond nearest neighbor interactions, finite barrier heights between sites, quantum mechanical, dynamic and exchange effects. Yet the experimental coefficients A are remarkably close to the theoretical values (Table 8.1), averaging just slightly greater than $\frac{1}{2}$ for both isotopes. Still further complications are apparent from the nature of the films at temperatures above the critical region. In Fig. 8.5 it may be seen that the heat capacity tends to the value Nk_B characteristic of a 2D gas rather than a monotonic decrease to 0 as in the localized model. From this it was suspected that the transition involves not only a spatial ordering but also a collective change in mobility, i.e., that the film above T_c is a disordered mobile monolayer while below T_c it is ordered and localized. This suspicion was confirmed by Rollefson (38) in a study of ^{3}He by nuclear magnetic resonance. Using cw techniques, Rollefson studied the resonance line width in the density–temperature region of the heat capacity peaks. He found substantially greater widths below T_c (consistent with dipolar broadening of localized adatoms) than above T_c (indicating motional narrowing); the lines were also narrow at higher and lower densities than the critical density, even at temperatures well below T_c. These results together with the heat capacity measurements, suggest a combined transition of spatial order and mobility, perhaps analogous to the "Mott transition" (39) between metallic and insulating states in 3D electronic solids (40).

Several theoretical studies have been specifically directed toward understanding the registered phase of helium on graphite and the nature of the order–disorder transition. Campbell and Schick (25) modeled the system as an array of localized classical atoms on a triangular lattice. The adatom interactions were assumed to be infinitely repulsive for atoms on the same site, a large finite repulsion between atoms on nearest-neighbor sites, a weak attraction between second-nearest neighbors, and no interaction at longer range. These characteristics approximate the real interactions between He atoms at the site spacing of the graphite surface. The results, based upon the Bethe–Peierls approximation, gave a first-order phase transition, for coverages near $\frac{1}{3}$, between a high-temperature disordered phase and a low-temperature phase with long-range order. The maximum transition temperature, occurring at $x = \frac{1}{3}$, was found to be located at $T_c \simeq J/k$, where J is the nearest-neighbor interaction energy. The theoretical phase diagram has a finite density range at intermediate temperatures, with regions of two-phase equilibrium at both higher and lower density. The failure of the theory to yield the proper order of the phase transition seems to be a consequence of the approximation method, which is known to give first-order transitions for which exact solutions predict second-order transitions. An additional limitation of the model is the assumption of site localization, which is evidently not the situation in the real films near and above T_c. The localized model may be appropriate for strongly bound layers such as Xe and N_2 on graphite, but in He films quantum mechanical zero point motion within sites and tunneling between sites cannot be neglected *a priori*. These effects, together with the influence of particle statistics, were studied by Schick and Siddon (26). The principal motivation for their calculation was the experimental observation of a slight isotope shift of the transition temperature, T_c being about 0.08°K higher for ^{3}He. To simplify the calculation, the model system was a square array, and the interactions were infinite repulsions between atoms on the same site and finite nearest-neighbor attractions (or repulsions). The calculation showed that in the absence of nearest-neighbor interactions, the dominant quantum effect is a reduction of T_c (^{3}He) more than of T_c (^{4}He) due to site-to-site tunneling and enhanced delocalization of the lighter isotope. With finite interactions between sites the effect becomes reversed, for then the greater amplitude of in-site motion of ^{3}He produces a larger average interaction energy. The calculated isotope effect for reasonable values of the interaction parameters has the correct sign and order of magnitude. The stability of the registered phase with respect to a 2D liquid phase was studied by Novaco (41). Using empirical values of the He–He interaction and the He–graphite potential, Novaco compared the ground state energy of a localized $\sqrt{3}$ array with a disordered mobile film of the same density and found that the registered arrangement has a slightly lower energy. Also considered were two other epitaxial structures that had been proposed earlier (37) but which had later been attributed (42) to incomplete equilibrium. In concordance with the deductions of Elgin and Goodstein, the proposed structures were found to be unstable with respect to the 2D liquid.

8.3 PARTIALLY REGISTERED MONOLAYERS; SUPERLATTICE PHASES

In the lattice gas models the substrate consists of an array of adsorption sites separated by indefinitely high potential barriers. On this basis an adatom can be found only at a site and not at an intermediate position. This approximation breaks down in the case of mobile adsorption, where, as discussed in Chapter 5, the limited height of the potential barriers permits thermally stimulated hopping and quantum tunneling from site to site, which means that there is a finite probability for finding adatoms at positions between sites. Here, also, in the present context of relatively dense monolayers, we find that with more realistic substrate potentials having finite heights some adatoms can be found at off-site locations. These off-site atoms tend to be arranged in a regular sequence relative to the substrate structure, in arrangements that are sometimes called "partially epitaxial" or "superlattice" phases. Such structures result from a kind of interference between the periodicities of the substrate potential and the preferred structure that the monolayer would have on a perfectly smooth surface. Superlattice phases ought to be the most common and typical states of films at low temperatures, as indicated by the following qualitative argument.

Let us first imagine a single-phase monolayer of arbitrary density on a uniform plane surface. At sufficiently low temperatures, the common phase will be a 2D solid, with some regular crystalline structure determined by the adatom interactions. We now switch on the periodic part of the substrate potential, gradually increasing the amplitude while keeping the symmetry and spacing at their final values. As the potential modulation grows the monolayer atoms are pushed in one direction or another, away from the barrier tops and toward the centers of the wells. Those that are closest to the barrier tops and site centers move the least, and those at intermediate positions, where the potential gradient is large, move more. But generally all adatoms are shifted to some extent, from their original equilibrium positions in the 2D array to more preferable locations when both adatom and substrate forces coexist. This readjustment of the monolayer produces an arrangement which is a composite of the original 2D crystal and the substrate structure. Even a weak substrate periodicity imposes its pattern on a film, for every real monolayer has a finite 2D compressibility, and every adatom can be shifted to some extent, even by the application of a very small perturbing force. Arguing in another way, if we begin first at the strong substrate limit, where the film is initially localized in adsorption sites and then switch on interactions between atoms in adjacent sites, a superlattice structure will develop owing to the finite restoring forces at each site. That is, except for unusual coincidences the equilibrium positions will be pulled away from the centers of the sites by interactions with the surrounding atoms, leading to a more preferable arrangement having a more complex symmetry than the original lattice gas array. Simple illustrations of these qualitative arguments are given in Fig. 8.6.

We now inspect the stability conditions for the various superlattice phases. Let us return to the first way of imagining the superlattice structure to develop, beginning from a purely 2D solid array and then switching on the substrate potential. Suppose we imagine this happening for two different monolayer densities, but keeping all of the other parameters the same. These two films will tend to develop different structures. At one density it might be most favorable to have a periodicity of, say seventeen adatoms and three sites, while at the other it would be better to have a different structure, perhaps one that repeats every fifty atoms and nine sites. The number of regular arrangements for a finite system is virtually as large as the number of atoms. Increasing the density of the monolayer by adding one atom will create a different starting condition leading to a new superlattice structure. The energies of the different structures may change by a large amount from one phase to the next. For example, if the number of atoms is

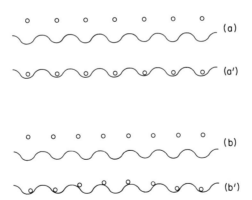

FIG. 8.6   Illustrations of simple and complex superlattice phases. (a) A regular atomic arrangement, with inherent preference for spacing coincident with the substrate spacing, leading to a simple epitaxial arrangement (a'). (b) The preferred spacing is not identical with the substrate periodicity, leading to the superlattice structure (b').

exactly equal to the number of sites the film can be in perfect registry, but with the addition or subtraction of just one atom the perfection is destroyed. The film would have to accept a much higher energy; an increase in total energy of $\sim \frac{1}{2}Nu$ in response to $\Delta N = \pm 1$. Surely the system can adjust to the new situation at lower cost, without such an increase in energy (at finite temperature, Helmholtz free energy). It can indeed, by breaking up into two phases of distinctly different densities. The situation is analogous to the two-phase equilibrium between states of ordinary matter, where a system at intermediate density achieves its lowest free energy by dividing into a mixture of a higher density and a lower density phase. Thus, nearly all of the superlattice structures that can be imagined will not actually occur; only a small number of especially favorable structures can persist. A qualitative illustration of the free energy diagram is shown in Fig. 8.7. The phases that will survive have free energies that are extrema on the coarse-

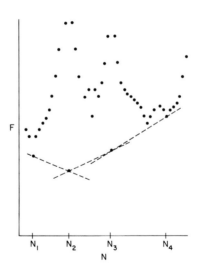

FIG. 8.7   Schematic diagram of the variation of Helmholtz free energy with the total number N of adsorbed atoms on a substrate of fixed area. Each dot represents the value of F of a pure single phase having the particular value of N. The dashed lines give the (lower) free energy of two-phase mixtures. Thus, in the region illustrated, only those structures which are the termini of the line segments could be observed.

grained free energy diagram; as depicted in Fig. 8.7, their free energies form the vertices of a segmented figure that is convex to the density axis from $N = 0$ to full coverage. Since the free energy is a function of T, the detailed shape of the figure will in general change with temperature, so that individual structures may have quite limited ranges of existence. There is no general way of predicting the number or symmetries of these phases, for they are resultants of many interactions, some of which may be in quite delicate balance.

There are several methods for detecting superlattice phases. The most direct are diffraction techniques; LEED, atomic beam diffraction, and neutron diffraction. LEED has been the most extensively used and it has yielded a great variety of structures, one of which is illustrated in Fig. 8.8. Atomic and neutron diffraction, while not in wide use, may have particular advantages over LEED in certain situations; investigations with these techniques may increase in the next few years.

It is also possible to deduce registry from calorimetric measurements. Ying (43), and Stewart and Dash (44) have shown that registry with the substrate causes an energy gap in the density of states of the film, eliminating all low lying vibrational modes below some finite frequency cutoff. The value of the cutoff frequency depends upon

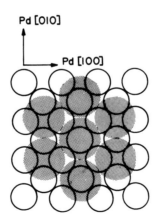

FIG. 8.8 An example of a partially registered phase: Xe on (100) Pd crystal surface, observed by Palmberg (32) via LEED.

the number of atoms in the repeating unit cell of the superlattice structure; what Ying calls the "order" of the arrangement (43). As a general rule the cutoff frequency decreases as the order increases. The appearance of the energy gap and its trend with the repeat distance of the structure can be seen in the following qualitative treatment (44).

We first consider a completely epitaxial monolayer in which each adatom in its ground state lies at the center of an adsorption site. For small deviations from equilibrium we can expand the potential energy in a power series in the displacements from the centers of the neighboring adatoms as well as displacements from the centers of the sites. Since the range of adatom interactions is typically short, the problem can be simplified by considering only the immediate vicinity of each adatom, e.g., only first and second nearest neighbors. Neglecting higher terms in the power series, one then obtains a series of dynamical equations similar to those for 2D lattices (45), but with the addition of extra terms due to adatom-site interactions. Wave solutions in the low k region yield a dispersion relation

$$\omega = [\omega_i{}^2 + c^2 k^2]^{1/2}, \tag{8.3.1}$$

where ω_i is the lateral frequency of isolated adatoms and c is the phase velocity of long wavelength "sound." There are no states of excitation $\omega < \omega_i$. The density of states rises abruptly from 0 at $\omega = \omega_i$ and then begins to follow the linear dependence of a 2D solid. For a square lattice of side d the density of states in the region of low frequency is

$$\begin{aligned} g(\omega) &= 0, & \omega &< \omega_i, \\ &= N d^2 \omega / 2\pi c^2, & \omega &> \omega_i. \end{aligned} \tag{8.3.2}$$

We next consider a partially epitaxial structure, where some atoms are not centered at sites. Assume that the substrate potential along one direction is $u(x)$ and that the film has some density distribution $g(X + x)$, where X is the center-of-mass coordinate of a small region of the film ("small" relative to the wavelengths of interest; since we are interested mainly in the region near $k = 0$, these regions may involve many atoms). Now low k excitations correspond to translations of these regions, the film structure being constant. The potential energy variation of the region extending from 0 to L is

$$U(X) = \int_0^L u(x)g(X + x)\, dx. \tag{8.3.3}$$

The restoring force is dU/dX; since g is symmetric between X and x,

$$dU/dX = \int u(\partial g/\partial X)\, dx = \int u(\partial g/\partial x)\, dx. \tag{8.3.4}$$

Differentiating once more and integrating by parts, we have

$$\frac{d^2U}{dX^2} = u(x)\left.\frac{\partial g}{\partial x}\right|_0^L - g\left.\frac{du}{dx}\right|_0^L + \int_0^L g\,\frac{d^2u}{dx^2}\, dx. \tag{8.3.5}$$

If L is large the first two terms are unimportant. The remaining term is just the total of the lateral force constants acting on the individual atoms near their equilibrium positions. If g and u have variations identical in periodicity and phase, then g and d^2u/dx^2 will be similarly in registry. The center-of-mass force constant will be equal to the sum of all of the constituent force constants, and since the mass is similarly scaled, the resonant frequency is equal to that of isolated adatom. But if g and u differ in periodicity and/or phase the potential $U(X)$ and the force constant d^2U/dX^2 will be reduced; hence the frequency of the mode at $k = 0$ will be some $\bar{\omega}_i < \omega_i$. For a completely misregistered or incommensurate structure, U and d^2U/dX^2 vanish and $\bar{\omega}_i = 0$. The densities of states of completely registered, partially registered, and completely misregistered structures are shown schematically in Fig. 8.9.

If there is a finite cutoff to the density of states, i.e., if $\bar{\omega}_i$ is nonzero, then the heat capacity of the film is exponential at low temperatures. At very low T ($\ll \hbar\bar{\omega}_i/k_B$), the heat capacity is given by

$$C = (2Nk_B/\pi)(d^2\bar{\omega}_i/c)(\hbar\bar{\omega}_i/k_BT)\,\exp\,(-\hbar\bar{\omega}_i/k_BT). \tag{8.3.6}$$

Two recent experimental studies have given evidence of superlattice phases through their low temperature exponential heat capacities. In Ar and Ne monolayers on Cu, the low T behavior was exponential, with a characteristic "Einstein" temperature much lower than could be expected from purely adatom–adatom or adatom–substrate interactions (42). It was concluded that the characteristic temperature was lowered due to an averaging process, i.e., by the fact of a long super-

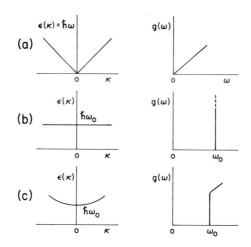

FIG. 8.9 Energy-momentum dispersion curves $\epsilon(k)$ and densities of states of surface–parallel vibrational modes of adsorbed films. (a) Smooth substrate or misregistered (incommensurate) solid monolayers. (b) Noninteracting localized atoms in harmonic adsorption sites. (c) Partially registered monolayer, with both substrate and interatomic forces.

lattice period causing $\bar{\omega}_i$ to be considerably reduced below ω_i. In a calorimetric study of Ne on graphite (46), similar exponential behavior led to an estimate of an effective characteristic temperature of $\sim 7°K$, again well below those that characterize either the substrate wells or the adatom interaction. At higher temperatures the heat capacity tended to the T^2 dependence of a 2D Debye solid.

REFERENCES

1. E. Wood, *J. Appl. Phys.* **35**, 1306 (1964).

2. J. J. Lander and J. A. Morrison, *Surface Sci.* **6**, 1 (1967).

3. P. B. Hirsh *et al.*, "Electron Microscopy of Thin Crystals," p. 317. Butterworths, London 1965.

4. R. H. Fowler, *Proc. Cambridge Phil. Soc.* **32**, 144 (1936).

5. R. H. Fowler and E. A. Guggenheim, "Statistical Thermodynamics," Chapter 10. Cambridge Univ. Press, London and New York, 1939.

6. R. E. Peierls, *Proc. Cambridge Phil. Soc.* **32**, 477 (1936).

7. H. A. Bethe, *Proc. Roy. Soc.* **A150,** 552 (1935).

8. G. G. Newell and E. W. Montroll, *Rev. Mod. Phys.* **25,** 353 (1953).

9. C. Domb, *Advan. Phys.* **9,** 149, 245 (1960).

10. M. E. Fisher, *Rep. Progr. Phys. (London)* **30,** 615 (1967).

11. T. D. Lee and C. N. Yang, *Phys. Rev.* **87,** 410 (1952).

12. L. Onsager, *Phys. Rev.* **65,** 117 (1944).

13. B. Kaufman, *Phys. Rev.* **76,** 1232 (1949).

14. R. M. F. Houtappel, *Physica* **16,** 425 (1950).

15. A. E. Ferdinand and M. E. Fisher, *Phys. Rev.* **185,** 832 (1969).

16. S. T. Wu, Ph.D. thesis, Univ. of California, San Diego, 1972.

17. L. P. Kadanoff *et al., Rev. Mod. Phys.* **39,** 395 (1967).

18. M. E. Fisher, *Proc. Roy. Soc. (London)* **A254,** 66 (1960).

19. B. M. McCoy and T. T. Wu, *Phys. Rev.* **176,** 631 (1968).

20. D. S. Gaunt and M. E. Fisher, *J. Chem. Phys.* **43,** 2840 (1965).

21. D. S. Gaunt, *J. Chem. Phys.* **46,** 3237 (1967).

22. W. W. Wood and J. D. Jacobson, *J. Chem. Phys.* **27,** 1207 (1957).

23. B. J. Alder and T. E. Wainwright, *J. Chem. Phys.* **33,** 1439 (1960).

24. C. K. Hall and G. Stell, *Phys. Rev.* **A7,** 1679 (1973).

25. C. E. Campbell and M. Schick, *Phys. Rev.* **A5,** 1919 (1972).

26. M. Schick and R. L. Siddon, *Phys. Rev.* **A8,** 339 (1973).

27. H. H. Farrell, M. Strongin, and J. M. Dickey, *Phys. Rev.* **B6,** 4703 (1972).

28. A. R. Ubbelhode and F. A. Lewis, "Graphite and Its Crystal Compounds." Oxford Univ. Press, London and New York, 1960.

29. R. M. Barrer, in "Non-Stoichiometric Compounds" (L. Mandelcorn, ed.), Chapter 6. Academic Press, New York 1964.

30. J. C. Tracy and P. W. Palmberg, *J. Chem. Phys.* **51,** 4852 (1969); *Surface Sci.* **14,** 279 (1969).

31. J. Suzanne, J. P. Coulomb, and M. Bienfait, *Surface Sci.* **40,** 414 (1973).

32. P. W. Palmberg, *Surface Sci.* **25,** 598 (1971).

33. J. C. Tracy, *J. Chem. Phys.* **56,** 2736 (1972).

34. J. C. Tracy, *J. Chem. Phys.* **56,** 2748 (1972).

35. J. Kjems, L. Passell, H. Taub, and J. G. Dash, *Phys. Rev. Lett.* **32,** 724 (1974).

36. B. F. Mason and B. R. Williams, *J. Chem. Phys.* **56,** 1895 (1972).

37. M. Bretz, J. G. Dash, D. C. Hickernell, E. O. McLean, and E. O. Vilches, *Phys. Rev.* **A8,** 1589 (1973).

38. R. Rollefson, *Phys. Rev. Lett.* **29,** 410 (1972).

39. N. F. Mott, *Proc. Phys. Soc. (London)* **62,** 416 (1949).

40. *Rev. Mod. Phys.* **40,** No. 4 (1968).

41. A. D. Novaco, *Phys. Rev.* **A7**, 1653 (1973).

42. R. L. Elgin and D. L. Goodstein, *Phys. Rev.* **A9**, 2657 (1974).

43. S. C. Ying, *Phys. Rev.* **B3**, 4160 (1971).

44. G. A. Stewart and J. G. Dash, *J. Low Temp. Phys.* **5**, 1 (1971).

45. A. A. Maradudin, I. P. Ipatova, E. W. Montroll, and G. H. Weiss, "Theory of Lattice Dynamics in the Harmonic Approximation," 2nd ed. (*Solid State Phys.*, Suppl. 3). Academic Press, New York, 1971.

46. G. B. Huff and J. G. Dash, *Proc. Int. Conf. Low Temp. Phys., 13th, Boulder, Colorado 1972* (K. D. Timmerhaus, W. J. O'Sullivan, and E. F. Hammel, eds.). Plenum Press, New York, 1974.

9. Heterogeneous Films

9.1 INTRODUCTION

All real substrates are heterogeneous to some extent, their adsorptive properties varying with position due to the several mechanisms discussed in Chapter 3. Heterogeneity of substrates affects all of the properties of their adsorbed films. The changes can be so great as to make the films qualitatively different from those on relatively uniform surfaces. Since heterogeneity is both powerful and ubiquitous, it has to be treated as an important intrinsic property and given consideration together with the heat of adsorption, surface area, substrate structure, and a few others that may be of controlling importance in various film regimes. In this chapter we take up a succession of regimes, comparing theory and experiment wherever possible.

There is a long history of theoretical interest in the subject, reaching at least as far back as Langmuir's work (1), but major questions remain unsolved. Langmuir sought to apply his newly developed microscopic theory of localized, sitewise adsorption to a model of a nonuniform surface; a composite of individually uniform sections or patches, each patch possessing a characteristic binding energy for gas molecules. It was assumed that adsorption on each type of patch obeyed the law derived for noninteracting localized atoms, and therefore the vapor pressure isotherm for the entire system could be ex-

pressed as the superposition of ideal Langmuir isotherms. Since the
adsorption parameters of each uniform patch would be just the number of sites available and the binding energy on each site of the patch,
the total heterogeneous surface could be characterized by specifying
the number of sites having different binding energies. In the general
case the number of sites is approximated as a quasi-continuous function of energy, as a density of states function $g(\epsilon)$. Thus the vapor
pressure isotherm for the entire heterogeneous surface can be expressed
as an integral of Langmuir isotherms, weighted by $g(\epsilon)$.

Langmuir's theory has stimulated a succession of papers continuing
to the present time; attempts to introduce greater physical realism to
the microscopic model, or to bring better agreement with experiment,
or to investigate the mathematical problem of inversion to extract $g(\epsilon)$
from the isotherm integral (2–8).

But there are also other schools of thought which reject one or more
of the basic postulates of that model, and begin with alternative assumptions about the form of the substrate heterogeneity, the nature
of the film, or both. Ross and Olivier (9) assume the film to be a
classical 2D van der Waals gas, but retain the notion that the substrate is composed of uniform patches ("homotattic"). Their model
leads, as does Langmuir's, to the conclusion that the vapor pressure
isotherm is an integral of uniform isotherms, modulated by a density
of states function for the distribution of binding energies.

The principle of superposition follows naturally from the assumptions of the Langmuir model, but it is by no means universal in all
theories. Hill (10), for example, studied a model heterogeneous surface
composed of a random spatial distribution of energetically different
sites, and he showed that for an interacting monolayer in the quasi-
chemical approximation, there could be two distinct regions of phase
condensation; first on the more preferential sites, and then at lower
temperatures on less favorable regions. Roy and Halsey (11), on the
other hand, argued that there ought to be some spatial correlation
between sites of similar binding energy, although not necessarily as
strongly correlated as in the homotattic model. In this event the binding energy variations could be nearly continuous functions of position,
producing lateral fields acting on the adatoms. With sufficiently strong
fields, a film might suffer a qualitative change in character, and there-

fore the inhomogeneous film could not be described as a simple superposition of uniform regions (12). These lateral fields need not necessarily be very large in order to cause dramatic effects in special cases.
For example, as Campbell *et al.* (13) pointed out, a 2D ideal Bose gas
is extremely sensitive to slight perturbing fields, which cause the film
to undergo a higher-order phase transition with a pronounced peak in
the heat capacity. Further studies of quantum monolayers on heterogeneous surfaces, by Widom and Sokoloff (14), Novaco (15), and Imry
and Gunther (16), concurred in demonstrating that pronounced qualitative changes could be induced in certain regimes, effects which could
not be represented by superposition.

  In addition to these schools of thought there are those which are
more empirical in their approach, primarily attempting to understand
the functional dependence of vapor pressure isotherms in real adsorption systems, using these as clues to deduce the distribution function
of binding energy and thereby the actual nature of heterogeneity. An
important line of investigation was begun by Dubinin and Radushkevich (17), who proposed a particularly successful empirical isotherm
equation for strongly heterogeneous surfaces. Subsequent experiments
have shown that the DR formula works well for a variety of gases and
surfaces over a very wide range of low coverages (18–25), and these
results (particularly those of Hobson and collaborators), together with
the insecure theoretical basis of the original paper, have led to several
attempts to put the DR equation on a firmer foundation. Cerofolini (7)
and Heer (26) have both offered derivations based on the localized site
model, but differing markedly in other respects. The problem of determining $g(\epsilon)$ from experimental vapor pressure or calorimetric data is
treated in detail in the next section. We find that, in the case of relatively strong heterogeneity and low temperatures, $g(\epsilon)$ is readily obtained without any need to assume that the adatoms are localized on
adsorption sites. With this general result we examine the significance
of the DR equation and other empirical results.

  Multilayer films have been studied with comparable intensity both
experimentally and theoretically. There are several theoretical vapor
pressure equations, the most famous being the Brunauer–Emmett–
Teller (BET) isotherm (27, 28). Figure 9.1 shows that it is successful
in describing many experimental curves (by means of three empirical

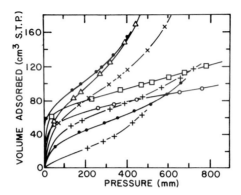

FIG. 9.1 Examples of vapor pressure isotherms having the sigmoid shape described by the equation of Brunauer, Emmett, and Teller (BET) (27). The smoothness of these isotherms, which is typical of most adsorbents, is due to a combination of substrate heterogeneity and relatively high temperature. Although the BET theory incorrectly attributes the smoothness to statistical variations in film thickness, the BET equation remains useful for estimating surface areas of common adsorbents. The figure is reprinted from the original article (27).

parameters) over the region from about one completed layer to the formation of bulk liquid. But in spite of its practical success, the theoretical basis of the BET law is internally inconsistent and unphysical; it assumes, for example, that each adsorbed atom interacts with those in adjacent layers but not with others in the same layer. It is now generally understood that the smooth sigmoid shape that is so typical of experimental isotherms is due, not to the actual conditions of the BET model, but to substrate heterogeneity. On very uniform surfaces at relatively low temperatures the vapor pressure exhibits a series of distinct steps with increasing quantity of gas adsorbed, each step corresponding to the completion of a layer. Stepwise isotherms have been obtained for Kr and other heavy rare gases adsorbed on graphite surfaces (both graphitized carbon black and exfoliated graphite) by several groups of investigators, particularly Halsey and his collaborators (29) (see Fig. 9.2) and by Thomy and Duval (30) (Fig. 9.3). Comparably sharp steps have been demonstrated by Larher (31), using a series of specially prepared adsorbents of divalent metal–halide layer compounds. Stepwise isotherms are readily explained as due

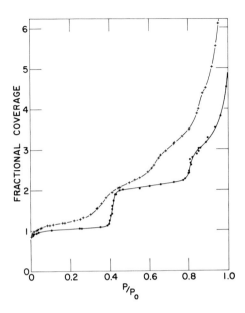

FIG. 9.2   Vapor pressure isotherms of argon (+) and krypton (●) on graphitized carbon black at 77°K, obtained by Singleton and Halsey (29). Since the
difference in energy between adjacent layers is smaller for Ar than Kr, thermal
promotion between layers is more significant in the Ar isotherm, causing the
steps to be much more rounded.

to the variation in binding energy with distance to the substrate,
and the experimental locations of steps scale properly according to the
principal theories of the adatom–substrate interaction (30). Since the
step locations are controlled by the binding energy to the substrate
(among other factors), on heterogeneous surfaces there will be an
assortment of steps at different pressures due to the completion of
layers on various portions of the substrate. With increasing heterogeneity, i.e., a wider distribution function of binding energy, there is
a progressive loss of detail, eventually resulting in a smooth isotherm
curve. This smooth superposition can be fitted by the BET isotherm.

   The historical emphasis of theoretical studies of heterogeneity has
been on its effects on vapor pressure behavior, a natural result of the
preponderance of experimental studies of this sort. But this emphasis
has retarded an examination of other film properties, some of which

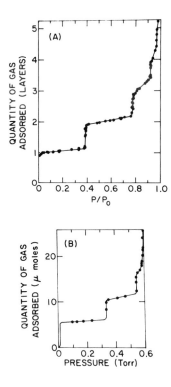

FIG. 9.3 Examples of stepwise vapor pressure isotherms on relatively uniform surfaces at low temperature. (A) Kr on exfoliated graphite at 77.3°K, measured by Thomy and Duval (30). (B) Kr on crystals of $CdBr_2$ at 73.1°K, measured by Larher (31).

might be more sensitive indicators of the magnitude and nature of substrate variations. It is unarguable that, no matter how carefully a thermodynamic property such as vapor pressure can be measured, it cannot yield a microscopic picture of the system except via its resemblance to assumed theoretical models. Certainly, there now are available a large number of more modern experimental techniques which offer high resolution analyses of certain substrate properties, and some of these may begin to answer long standing questions about physisorption. One of the most persistent is the question of spatial distribution of heterogeneity, e.g., are real surfaces "homotattic," or are there long-range variations which act as lateral fields? More generally, what is the local distribution $g(\mathbf{r}, \epsilon)$ of physisorption states or sites in both positional and energy space? Ideally, what is needed is a kind of topographical map of binding energy over a surface, where both the locations and values of the stronger and weaker binding regions can be seen together. Films can themselves be used as probes for giving such detail. An example of such an application is described in Sec. 9.6.

9.2 VAPOR PRESSURE ISOTHERMS AND ISOSTERES AT LOW COVERAGES

It has been shown in Chapter 5 that the vapor pressure isotherms of all gases on uniform substrates must tend to Henry's law at sufficiently low coverages, independent of the temperature, mobility, and interactions. "Sufficiently low coverage" will vary from system to system and may in particular cases be difficult to study, but nevertheless Henry's law must obtain in the limit. However, in most systems the effects of heterogeneity intervene before the coverage is reduced to the linear region. In these circumstances it turns out that the vapor pressure isotherm can be used to survey the variations of binding energy on the surface, as we show by the following simple theory.

Assume a finite quantity of adsorbent with monolayer capacity N_m of a certain gas. The gas molecules have adsorption energies ϵ which vary with position along the surface. The energy variation can be described in terms of a distribution function $g(\epsilon)$ such that

$$\int_{-\epsilon_{\max}}^{-\epsilon_{\min}} g(\epsilon)\, d\epsilon = N_m. \tag{9.2.1}$$

The finite monolayer capacity derives from the short-range repulsive interactions between molecules, which prevent more than one particle from occupying the same site or region of the surface. This "hard-core exclusion" is analogous to the Pauli exclusion of Fermi–Dirac particles, and it causes the adsorbed particles to have the same form of occupation probability $f(\epsilon)$ (32)

$$f(\epsilon) = [e^{\beta(\epsilon-\mu)} + 1]^{-1}, \tag{9.2.2}$$

where μ is the chemical potential. Just as in the familiar case of electrons in metals, the occupation probability at low temperature tends toward a limiting step shape, all states with less than a certain energy being occupied and those above being empty. At higher temperatures the "Fermi edge" becomes less sharply defined. The total number of adsorbed particles satisfies the sum rule

$$N = \int_{-\epsilon_{max}}^{-\epsilon_{min}} f(\epsilon) g(\epsilon) \, d\epsilon \qquad (9.2.3)$$

irrespective of the temperature.

Since real adatoms have long-range interactions, the adsorption of more molecules changes the energies of particles already on the surface, hence $g(\epsilon)$ is itself a function of the total state of the monolayer. However, if we adopt the simplification which in the theory of metals is called the "rigid band approximation," then the theory becomes quite tractable. At relatively low temperatures $(\beta\mu \gg 1)$ we can expand the integral in Eq. (9.2.3) by the familiar asymptotic series (33), obtaining

$$N = \int_{-\epsilon_{max}}^{\mu} g(\epsilon) \, d\epsilon + (\pi^2/6)(kT)^2 (dg/d\epsilon)_{\epsilon=\mu} + \cdots \qquad (9.2.4)$$

Now we consider a small increase in coverage at constant T:

$$\dot{N} + \delta N = \int_{-\epsilon_{max}}^{\mu+\delta\mu} g(\epsilon) \, d\epsilon + (\pi^2/6)(kT)^2 \, \delta[(dg/d\epsilon)_{\epsilon=\mu}]. \qquad (9.2.5)$$

Taking the difference between Eqs. (9.2.5) and (9.2.4), we obtain to first order

$$\delta N = \int_{\mu}^{\mu+\delta\mu} g(\epsilon) \, d\epsilon = g(\mu) \, \delta\mu;$$

hence

$$(\partial N/\partial\mu)_T = g(\mu). \qquad (9.2.6)$$

The chemical potentials of film and vapor are equal in equilibrium. If the vapor is nearly ideal, $\mu_v = kT \ln (P\lambda^3/kT)$; then evaluating $(\partial N/\partial\mu_v)_T$, we obtain from Eq. (9.2.6)

$$(\partial \ln P/\partial N)_T = 1/kT g(\mu). \qquad (9.2.7)$$

Equation (9.2.7) is a prescription for mapping out the binding energy distribution directly from a vapor pressure isotherm. It was derived explicitly for an adsorbed film of noninteracting fermions (34) a few years ago, and is discussed in the context of surface band states in

Chapter 5. However, it was first obtained many years before by Roginskii (2), for heterogeneous adsorption. Roginskii's derivation is based upon the assumption of localized adsorption on sites, which we show here is unnecessarily restrictive. In addition, Roginskii's treatment is somewhat intuitive in part, which may explain why it has not received the attention it deserves.

Vapor pressure isosteres can also be used to map the distribution, although they require the added complication of a temperature sweep. Using the definition of the isosteric heat of adsorption (Chapter 4), we obtain directly from Eq. (9.2.7)

$$(\partial q_{st}/\partial N)_T = -1/g(\mu). \tag{9.2.8}$$

Equation (9.2.8) can be recognized as asymptotically correct at low T just as it stands without the preceding development. An equivalent relation was stated without proof by Drain and Morrison (35) and applied by them to an analysis of experimental heats of adsorption.

Equations (9.2.7) and (9.2.8) are quite general but their underlying assumptions must be kept in mind. In addition to the rigid band approximation, which is tantamount to a neglect of adatom interactions other than exclusion, the film is assumed to be at relatively low temperatures, $\beta\mu \gg 1$. This condition is effectively one of strong heterogeneity, so that the "Fermi edge" is sharply defined in comparison to μ.

Equation (9.2.8) is readily applied to analyze experimental isotherms to obtain the energy distributions of actual surfaces. Of the many empirical isotherm equations in the literature, there are two that have been found particularly successful in describing various systems, particularly at quite low submonolayer coverages; the Freundlich (36) and the Dubinin–Radushkevich (17) isotherms. The Freundlich isotherm is

$$N/N_m = (P/P_0)^{aT}, \tag{9.2.9}$$

where N/N_m is the fractional coverage at P, T. P_0 and a are usually taken as constants, although a is sometimes assigned a weak dependence on T (8). The DR isotherm has the quite different form of

$$N/N_m = \exp(-B\zeta^2), \tag{9.2.10}$$

where B is a constant and ζ is the "Polanyi potential" (37)

$$\zeta = RT \ln P/P_s,$$

P_s being the vapor pressure of the bulk adsorbate at T. Straightforward application of the general transformation equation (9.2.8) yields for the two isotherms:

Freundlich

$$g(\mu) = (a/k)N_m(kT/P_0\lambda^3)^{aT}e^{a\mu/k}; \qquad (9.2.11)$$

Dubinin–Radushkevich

$$g(\mu) = 2BN^2(\mu - \mu_0) \exp[-BN^2(\mu - \mu_0)^2], \qquad (9.2.12)$$

where we have used the standard ideal gas formula for $\mu(P)$. These two formally different distributions might imply quite different types of surfaces and mechanisms for heterogeneity, were it not for the fact that the two isotherms from which they were derived are actually quite similar. Although Eqs. (9.2.9) and (9.2.10) are completely distinct functions, they look alike over restricted ranges of coverage and pressure. The Freundlich isotherm describes a linear variation between $\log N$ and $\log P$, while the DR formula has $\log N$ varying as $(\log P)^2$; thus if the data do not extend over more than a few decades, either formula might give a reasonable fit. Since a traditional style of displaying data is on a log–log plot, it is natural that, as long as the vapor pressure is a fairly smooth function of P, agreement with the Freundlich equation should emerge. This is presumably the reason for the long popularity of that formula. But more recent work on a variety of systems has shown that, with extensions to lower pressures and coverages, curvature is invariably seen in the logarithmic plots, and the data are much more closely fitted by the DR law (20–25). See Figs. 9.4 and 9.5. Therefore it is the distribution function equation (9.2.12), dominantly gaussian in form, that appears to be typical of virtually all surfaces at low coverage. Before attempting to interpret it, we must clarify two points. The first is the appearance of P_s in the DR isotherm, for the vapor pressure of the bulk adsorbate can have nothing to do with a low coverage isotherm: indeed, the derivation of the general transform equation

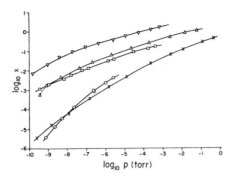

FIG. 9.4 Comparison of adsorption isotherms at 77.4°K for various gases on Zr (20) and Pyrex (21) as presented by Hobson and Chapman (25). The data can be described by the Freundlich (power law) isotherm over limited intervals, but all exhibit definite curvature. (∇) Xe on Zr; ($\triangle$) Kr on Zr; ($\square$) N_2 on Pyrex; ($\bigcirc$) Ar on Pyrex; ($\times$) Ar on Zr.

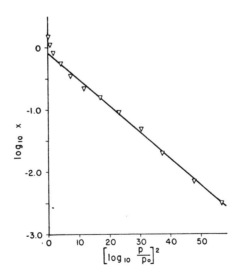

FIG. 9.5 Test of the Dubinin–Radushkevich isotherm equation for Xe on Zr at 77.3°K. The data were obtained by Hansen (20) and presented in the form shown here by Hobson and Chapman (25).

(9.2.7) involves only the adatom–substrate interaction and the effective size of the adatoms. In fact, it turns out that there is no experimental proof that the proper constant in the DR law is P_s and not some other number of comparable size. Moreover, slight deviations of wide-ranging data from the formula might be removed if P_s were treated as an adjustable parameter (25). The other supposedly predetermined parameter is N_m, but it turns out that, instead of being the actual monolayer capacity, it must be treated as an empirical constant in order to make the DR formula fit certain gas–surface combinations (25). Therefore, if both P_s and N_m are empirical, the DR law is seen to imply just a gaussian distribution of binding energy. We attempt to rationalize it as follows.

Very uniform surfaces have distribution functions which are sharply peaked at a single value of energy, ideally approaching a δ function shape. Less uniform surfaces must have less singular distributions which may take various forms. If the adsorbent is composed of a number of different species of surface (such as the different facets of a single type of solid crystal), each species being uniform, the distribution will consist of a series of sharp peaks. Other forms of heterogeneity can produce broader distributions either in combination with one or more sharp peaks or, in the case of a highly heterogeneous surface, a broad curve without any distinctive portions, perhaps resembling a gaussian. Indeed, we must expect gaussian character in at least portions of all real distributions, particularly at the extremes of binding energy. Even for highly uniform surfaces there will be very small fractions of unusually strong and also unusually weak binding regions, due to strains, cracks and all types of imperfections. Since the imperfections are of different origins and are to some degree statistically independent, their combined effects on binding energies tend to be distributed in a gaussian curve:

$$g(\epsilon) = C \exp\left[-\alpha(\epsilon - \epsilon_0)^2\right]. \tag{9.2.13}$$

Now if the entire surface followed such a distribution the coefficient C would be related directly to N_m and α, through Eq. (9.2.1). This is not usually the case, even for a highly heterogeneous surface, for there may be a maximum in $g(\epsilon)$ which is more sharply peaked than a simple gaussian. Therefore the Eq. (9.2.13) will generally be true only in the wings of the distribution, and the parameters C, α, and ϵ_0 do not necessarily have any physical significance outside of the wings.

There is still another controversial point concerning low coverage isotherms, having to do with the appearance of Henry's law. It has been stated that the limiting behavior of every isotherm at sufficiently low pressures must be Henry's law, i.e., characteristic of adsorption on the patches of surface of strongest binding (38). This claim is doubtful because it can be argued that the strongest binding region need not be a finite area patch but only a single site, at a confluence of a large number of independent features, or so few in number that the vapor pressures at which they adsorb are virtually inaccessible. Indeed, vapor pressure isotherms followed to very low pressures have shown no sign of evolving from Eq. (9.2.10) toward Henry's law (25). On the contrary, the linear region is found, if it exists at all, in the region of moderately low to medium coverage, after all of the especially energetic regions are covered. An especially clear example of a progression from heterogeneity at low coverage to uniformity at intermediate coverage is seen in the empirical distribution of ^{4}He adsorption energies on Grafoil [see Eq. (3.4.1), p. 54].

9.3 HEAT CAPACITY OF STRONGLY HETEROGENEOUS MONOLAYERS

Heat capacities of adsorbed films can probe the binding energy distribution, yielding the same quantity $g(\epsilon)$ that controls the vapor pressure isotherms and isosteres. On the basis of the simple "fermion" model of the previous section, we obtain the low temperature expression for the film energy as an asymptotic series similar to Eq. (9.2.4) (33), i.e.,

$$E = \int_{-\epsilon_{max}}^{\mu} \epsilon g(\epsilon)\, d\epsilon + (\pi^2/6)(kT)^2 [(d/d\epsilon)(\epsilon g(\epsilon))]_{\epsilon=\mu} + \cdots . \quad (9.3.1)$$

The heat capacity $C = dE/dT$ is given to first order by

$$C = (\pi^2/3)k^2 T g(\mu), \quad (9.3.2)$$

a formula identical to that for a fermion gas at very low temperatures (39). Equation (9.3.2) predicts that the heat capacity of a strongly heterogeneous film is linear in T at low temperatures, whatever the form of the distribution function. We emphasize that the result pertains to strong heterogeneity by noting the disparity between these results and those of other models in which the heterogeneity is quite weak, discussed later in this chapter. The use of Eq. (9.3.2) for probing surface heterogeneity is analogous to the exploration of the density of electron states in a metal by studying the changes in the low temperature heat capacity produced by alloying (40).

To the present order of approximation the density of states $g(\mu)$ determined by calorimetric measurements is the same as that obtained from vapor pressures. Therefore Eq. (9.3.2) can be combined with (9.2.7) or (9.2.8) to give a direct cross connection between these two types of measurement on a strongly heterogeneous surface:

$$C = -(\pi^2/3)k^2T(\partial q_{st}/\partial N)_T^{-1}. \qquad (9.3.3)$$

A combined vapor pressure and calorimetric study of the same system could therefore test the consistency of the theory and test the validity of the simplifying assumptions. We are aware of only one such test; Princehouse's study of ^{4}He adsorption on sintered Cu at temperatures 1–10°K (41). Examining his data over the range of low and intermediate coverages, we find that Eq. (9.3.3) is approximately obeyed, the heat capacity data yielding about 30% higher $g(\mu)$. This order of agreement seems reasonable in view of the simplicity of the theory.

9.4 WEAK HETEROGENEITY: EFFECTS ON LOW COVERAGE FILMS

The simple approximation of the previous section breaks down when the adatoms can no longer be treated as hard spheres without thermal excitation. Here we consider the opposite limit of weak heterogeneity, beginning first with the simple model of a 2D Boltzmann gas, i.e., of a sufficiently dilute monolayer at relatively high temperatures. We can easily modify the Helmholtz free energy of a uniform ideal 2D gas monolayer derived in Chapter 5. The substrate heterogeneity is introduced in the form of a spatially dependent perturbation $u'(\mathbf{r})$ in the adatom–substrate interaction, so that the single particle energy function is changed from $(p^2/2m - \epsilon_0)$ to $(p^2/2m - \epsilon_0 + u'(\mathbf{r}))$. Carrying through the usual steps and defining F_0 as the unperturbed free energy, we obtain for the heterogeneous film

$$F = F_0 - NkT \ln \left\{ (1/A) \int \exp\left[-\beta u'(\mathbf{r})\right] d^2r \right\}. \qquad (9.4.1)$$

The Boltzmann approximation is appropriate at relatively high temperatures and low densities. For high temperatures ($\beta u' \ll 1$) we can treat the correction term by thermodynamic perturbation theory (33), first expanding the exponential and then the logarithm to second order in $\beta u'$ to obtain

$$F = F_0 + N\overline{u'} + (N/2kT)[(\overline{u'})^2 - \overline{(u')^2}], \qquad (9.4.2)$$

where the bar over a quantity represents an areal average, e.g.,

$$\overline{u'} = A^{-1} \int u'(r)\, d^2r. \qquad (9.4.3)$$

From the perturbed free energy we can derive the changes in the usual thermodynamic quantities. Two properties of interest are the isosteric heat of adsorption and the heat capacity, which are, to second order

$$q_{st} = (q_{st})_0 + \overline{u'}, \qquad (9.4.4)$$

$$C = C_0 + (N/kT^2)[\overline{(u')^2} - (\overline{u'})^2]. \qquad (9.4.5)$$

Thus the effect on q_{st} is simply a shift in value with no temperature dependence, while the heat capacity receives a new and distinctive T^{-2} term. This type of temperature dependence is seen at relatively high temperatures in all sorts of systems having a cluster of low lying energy states, and it appears here as a general result for any form of weak heterogeneity. In the particularly simple case of a linear variation of binding along the surface we obtain

$$q_{st} = (q_{st})_0 + \Delta/2, \qquad (9.4.6)$$

$$C = C_0 + \tfrac{1}{12} NkT(\Delta/kT)^2, \qquad (9.4.7)$$

where Δ is the magnitude of the total change in binding energy.

9.5 QUANTUM GASES

We now explore the effects of weak substrate inhomogeneity on gaseous monolayers in the low temperature regime, where statistical interactions become important (13–16). Since a spatial variation of substrate properties lifts the translational degeneracy, we must explicitly expose the spatial dependence of the single-particle states of the film, and this can be accomplished by first writing particle number N and energy E in terms of the quasi-classical density of states, leaving explicit in the integrand the differential area d^2r (see Chapter 5)

$$N = \iint \frac{d^2p \, d^2r/h^2}{e^{\beta(\epsilon-\mu)} \pm 1}, \qquad (9.5.1)$$

$$E = \iint \frac{\epsilon \, d^2p \, d^2r/h^2}{e^{\beta(\epsilon-\mu)} \pm 1}. \qquad (9.5.2)$$

In the conventional theory $\epsilon = p^2/2m$ (we are ignoring the binding energy ϵ_0 here), but now we assume an additional term due to the variation in adsorption energy over the surface. Therefore we cannot simply integrate d^2r to obtain the area A as in the usual calculation, but now have to deal with double integrals of the Bose and Fermi functions. We can explore some of the interesting features of the model if the variation has a simple form. One such form is a linear variation, and we will outline the calculation for this case. Explicitly treating the substrate as having a limited extent, the total energy variation is some finite Δ, so that we can write

$$\epsilon = p^2/2m + f(x, y) = \epsilon' + bx, \qquad b = \Delta/X,$$

where

$$0 \lesssim x \lesssim X, \qquad 0 \lesssim y \lesssim Y, \qquad XY = A.$$

Then we obtain

$$N = \frac{2\pi m A}{h^2 X} \int_0^X dx \int_0^\infty \frac{d\epsilon'}{e^{\beta(\epsilon'+bx-\mu)} \pm 1}, \qquad (9.5.3)$$

$$E = \frac{2\pi m A}{h^2 X} \int_0^X dx \int_0^\infty \frac{(\epsilon' + bx) \, d\epsilon'}{e^{\beta(\epsilon'+bx-\mu)} \pm 1}. \qquad (9.5.4)$$

The Bose system is particularly interesting, for it exhibits a combined spatial and momentum condensation at a finite temperature. To demonstrate this, we compute the temperature T_0 at which $\mu = 0$. From Eq. (9.5.3) we have

$$N = \frac{2\pi m A}{h^2 X} \int_0^X dx \int_0^\infty \frac{d\epsilon'}{\exp\,[\beta_0(\epsilon' + bx)] - 1}, \qquad (9.5.5)$$

where $\beta_0 = (kT_0)^{-1}$. The integral over the kinetic energy ϵ can be obtained in closed form as

$$\int_0^\infty \frac{d\epsilon'}{\exp\,[\beta_0(\epsilon' + bx)] - 1} = \frac{1}{\beta_0} \ln\,[1 - \exp\,(-\beta_0 bx)].$$

If we assume that the total substrate variation is small, i.e., $\beta_0 \Delta \ll 1$, then the exponential can be expanded to first order as

$$\ln\,[1 - \exp\,(-\beta_0 bx)] \cong \ln\,(\beta_0 bx).$$

Then Eq. (9.5.5) becomes

$$N = -(2\pi m A/h^2\beta_0 X) \int_0^X dx\,\ln\,(\beta_0 bx) = (2\pi m A/h^2\beta_0)[1 - \ln\,\beta_0\Delta] \qquad (9.5.6)$$

Therefore, the condensation temperature is given by the transcendental equation

$$kT_0 = (h^2/2\pi m)(N/A)[1 - \ln\,(\Delta/kT_0)]^{-1}. \qquad (9.5.7)$$

Equation (9.5.7) has a form similar to T_0 for a 2D Bose gas on a uniform substrate of finite area, where in that case the argument of the logarithm is ϵ_m/kT_0, ϵ_m being the energy of the lowest translational state due to the finite dimensions of the film. In the present case as well as the finite area film, $T_0 \to 0$ as the energy parameter Δ or ϵ_m vanishes, thus recapturing the condition that in infinite uniform 2D geometry there can be no ordering at finite temperature. However, a finite T_0 does not necessarily imply any strong peaks or breaks in slope of the heat capacity; a sharp signal at T_0 is seen only when ϵ_m is itself appreciable in comparison to kT_0. For inhomogeneous films we find that a similar condition is required of the total inhomogeneity Δ. To explore the heat capacity near T_0 we need to solve the energy equation (9.5.4), and this can be done by means of a series expansion given by Robinson (42) and London (43) in connection with the 3D gas. The general Bose–Einstein integral is defined as

$$F_\sigma(\alpha) \equiv \frac{1}{\Gamma(\sigma)} \int_0^\infty \frac{y^{\sigma-1} \, dy}{e^{y+\alpha} - 1}, \qquad (9.5.8)$$

where Γ is the factorial function. Then the thermal energy E of the Bose monolayer can be written

$$E/K = (kT)^2 \int_0^X \Gamma(2)F_2(\alpha) \, dx - bkT \int_0^X \ln \, (1 - e^{-\alpha})x \, dx, \qquad (9.5.9)$$

where $K = 2\pi m A/h^2$ and $\alpha = \beta(bx - \mu)$. The expansion for $F_2(\alpha)$ is (43)

$$F_2(\alpha) = \frac{\pi^2}{6} + \alpha(\ln \alpha - 1) + \frac{\alpha^2}{4} + \frac{\alpha^3}{3!60} + \cdots. \qquad (9.5.10)$$

With this series it is easy to find the heat capacity at $T \gtrless T_0$ as long as $kT_0 \gg \Delta\epsilon_0$: after substitution and reduction one obtains

$$C(T \gtrless T_0) = \frac{2\pi^3 m A}{3h^2} k^2 T = \frac{\pi^2}{3} Nk \left[\frac{T/T_0}{1 - \ln \, (\Delta/kT_0)} \right]. \qquad (9.5.11)$$

Thus the heat capacity at $T < T_0$ varies linearly with the temperature, and if $\Delta/kT_0 \lesssim 0.1$ it rises to a peak value at T_0 greater than the classical value. A more complete expansion and numerical calculations show that $(dC/dT)_{T_0}$ is discontinuous, C_{T_0+} having an infinite negative slope. At higher temperatures the heat capacity falls rapidly with increasing T, approaching the T^{-2} variation of the classical regime at $T \lesssim 2T_0$. The numerical calculations show that C continues to display a cusp at T_0 even for much smaller values of Δ/kT_0, so that the onset of ordering is still detectable for quite minor substrate inhomogeneities (13). In contrast.to bosons, the theory predicts that fermions are much less sensitive to lateral fields, showing little change from the field-free behavior even for $\Delta/kT_0 \sim 1$. Comparisons between Bose and Fermi gases are presented in Fig. 9.6, as computed heat capacities for several values of Δ.

There is a history of theoretical interest in the heterogeneous Bose gas. Goldstein (44) explored the effects of a gravitational field on the 3D ideal Bose gas, and showed that it caused a spatial segrega-

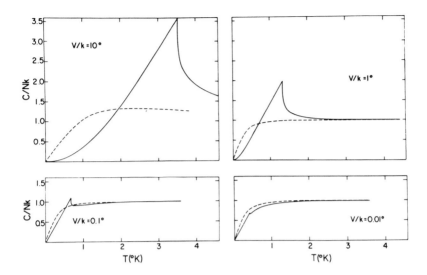

FIG. 9.6 Specific heats of two dimensional ideal Bose (———) and Fermi (– – – –) gases in weak lateral fields calculated by Campbell *et al.* (13). The results show that quite minor substrate heterogeneities may in some cases cause important qualitative changes in film properties.

tion along with momentum condensation. Lamb and Nordsieck (45) pointed out that the condensed phase in Goldstein's calculation is sharply localized in space at the position of lowest potential energy, and this unrealistic δ function in density was a result of the quasi-classical phase space approximation. If, instead, one considers the discrete character of the energy states in the gravitational field, one finds that the condensed phase is in the lowest quantum state of vertical motion, and that the thickness of the layer on the floor is of the order $t \simeq (h^2/m^2 g)^{1/3}$. For g equal to the standard value of the acceleration of gravity, the thickness of a condensate of an ideal mass 4 gas is about 5×10^{-3} cm. The effect of a gravitational or other external field is more important to the 2D gas, which would not otherwise undergo momentum condensation. Widom (46) specifically considered the effects of gravitational and centrifugal fields on the ideal 2D gas, and showed that moderate angular velocities could produce experimentally accessible condensation temperatures. The calculation was

criticized by Rehr and Mermin (47) for the unphysical divergence of
the condensate density. In addition to the inapplicability of the quasi-
classical approximation in this case, the ideal gas model must fail as
the density approaches hard core contact between the atoms. Nevertheless, some effects may be seen in the heat capacities of real monolayers at relatively high temperatures and low densities, where there
is no appreciable occupation of the lowest quantum state. Thus, although one cannot expect a true "Bose–Einstein condensation" in a
real film, it may be possible for some effects of lateral fields due to
substrate heterogeneity to become manifest in the heat capacities of
^{4}He monolayers at low temperatures.

9.6   EFFECTS OF HETEROGENEITY ON PHASE CHANGES IN MONOLAYERS

Regions of phase condensation are very sensitive to heterogeneity,
for it is in the neighborhood of transitions that thermodynamic systems (of any dimensionality) are especially "soft"; indeed, the compressibility becomes infinite at a first-order transition. It is just this
divergence in the volume susceptibility that leads to gravitational
separation of ordinary liquid and vapor or liquid and solid phases in
a vessel; otherwise the relatively weak gravitational energy is virtually undetectable. In an analogous manner we expect that weak lateral
fields due to substrate heterogeneity will tend to cause a two-phase
film to distribute itself unevenly on the surface, the dense phase going
to the regions of higher binding. This means that a two-phase monolayer could be used as a most delicate detector of substrate heterogeneity, the location and distribution of dense phase patches affording
a fairly direct "map" of the preferential regions. This mechanism is
operative in the Au-decoration technique for visualizing surface imperfections by electron microscopy (48). The Au atoms are evaporated
on to a heated substrate where they diffuse along the surface and tend
to congregate in regions of stronger binding where they nucleate into
clusters of 3D metal. When these clusters grow to sufficient size they
become visible in the microscope (see Fig. 9.7).

Effects of heterogeneity on a two-phase monolayer can also be observed in thermal measurements. Suppose we first imagine a mono-

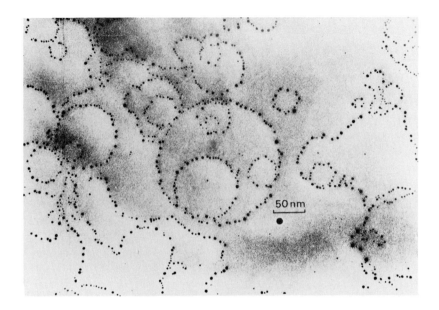

FIG. 9.7 Electron microphotograph of the surface of a graphite crystal using a Au-decoration technique to visualize surface heterogeneities (48). Vacuum deposited Au tends to nucleate in regions of higher binding energy. In this photograph the Au clusters outline etch patterns of ledges only 1 or 2 atoms thick. In other examples the authors obtain clustering at even weaker heterogeneities due to subsurface dislocations. (Photograph courtesy of J. M. Thomas.)

layer as a single phase, for example as a 2D gas above its critical temperature. Assume that the film has a long-range density variation due to heterogeneity or to incomplete equilibration in adsorption from the vapor. The system is now cooled so that 2D condensation begins. If the density were uniform it would begin condensing everywhere throughout the sample, producing a discontinuous jump in the total heat capacity such as the idealized Fig. 4.4 of Chapter 4. This jump occurs where the 2D vapor density $n(T)$ equals the average density n of the monolayer. However, if the actual monolayer density is some nonuniform function $n(\mathbf{r})$ then condensation occurs over a range of temperatures, and the total heat capacity signal is not a discontinuity but a rounded peak reflecting the spread in coverage. Here we can make a simple estimate of the blurring of the discontinuity in terms of the width of the coverage distribution. Assume that the 2D vapor density along the equilibrium line is a function of the form

$$n(T) = n_0 \exp (-q/kT), \tag{9.6.1}$$

which corresponds to a relatively low vapor density and constant 2D heat of vaporization q. Defining T' as the temperature of condensation when $n(T)$ equals the local coverage, we find for relatively small variations that the condensation anomaly has a temperature width $\delta T'$ proportional to the variation δn:

$$\delta T' \simeq (\overline{T'})^2 [q/k]^{-1}(\delta n/\bar{n}), \tag{9.6.2}$$

where $\overline{T'}$ is the condensation temperature for the average density $\bar{n}$.

It is this type of heterogeneity that is presumed responsible for the rounded heat capacity peaks seen in ^{4}He monolayers at low coverages near $1°$K and in Ne monolayers near $20°$K (see Figs. 6.2 and 7.5). Note that density variations of the type discussed do not blur or round a triple point singularity. For here one is dealing with the melting of patches of 2D solid in equilibrium with 2D vapor, and since the vapor pressure is independent of the size of the vapor or solid region (if not too small), the melting temperature is the same for all portions of the sample. Thus we understand why the Ne films have very sharp *melting* anomalies while their condensation peaks are quite diffuse.

A phase change induced by substrate heterogeneity is understood to be the mechanism for the persistence of 2D solidlike ("pseudo Debye" (11)) heat capacities of He monolayers on sintered copper sponge at low coverage and high temperature (12, 41, 49–52). Vycor glass surfaces evidently have comparable heterogeneity, for the heat capacities of He films on Vycor (53, 54) are quantitatively similar to those on copper sponge. The phenomenon is apparent even on Grafoil in He films at very low coverage. At fractional coverage $x \gtrsim 0.07$ and temperatures as high as $4°$K the heat capacities resemble those of 2D solids, whereas at somewhat higher coverage the films appear as 2D gases (55). The anomalous low-density behavior has been explained by Elgin and Goodstein (56) as small regions of dense film on the minor fraction of the substrate having abnormally high binding energy. Their model is in quantitative agreement with the observations, being based on the empirical distribution of binding energy and the thermodynamic functions of the intermediate density regimes.

9.7 MULTILAYER FILMS

Adsorption on an ideal uniform substrate at relatively low temperature produces a series of sharp steps due to the successive completion of individual layers. Two examples of such behavior in Kr adsorption, obtained by Thomy and Duval on exfoliated graphite (30), and by Larher on $CdBr_2$ (31) are shown in Fig. 9.3.

Although there have been several combinations of gases and substrates which have yielded stepwise isotherms, such behavior is by no means typical. The usual result obtained with an arbitrary substrate is a smooth signoid curve which can be represented fairly well by the so-called Brunauer–Emmett–Teller (BET) isotherm (27, 28). The BET law is a two-parameter equation between vapor pressure P and volume v of adsorbed gas:

$$x/v(1 - x) = (1/v_m c) + [(c - 1)/v_m c]x, \qquad (9.7.1)$$

where $x = P/P_0$, P_0 being the vapor pressure of the bulk adsorbate at the temperature of the isotherm. The constant v_m represents the volume of the first completed layer, while c is related to the free energy difference between the first and succeeding layers.

The BET equation is quite useful as an analytical tool for extracting certain gross features from empirical isotherms on heterogeneous surfaces, but it cannot be considered as having any fundamental significance. Its various derivations rest upon a highly unrealistic and internally inconsistent model which has been roundly criticized by Halsey (57, 58) and others. The model assumes that the first layer adsorbs as a Langmuir monolayer having some characteristic binding energy to the substrate, and that successive layers adsorb with the binding energy of liquid. The assumption that the substrate field is no longer felt at layers beyond the first is incorrect, but it is not the most serious defect. Much more unrealistic is the assumption that higher layers can be formed statistically by deposition on single ad-atoms, i.e., with no lateral interactions. The model thus removes any significance of liquid like properties while at the same time assuming that the liquid represents the state of all higher layers. But the correct explanation for the BET isotherm is that it is a superposition of stepwise isotherms from different portions of a heterogeneous surface, the steps occurring at different pressures on different facets or otherwise distinctive regions of the substrate. An additional factor tending to obscure steps even on a uniform surface is thermal excitation between layers, because the distinction between one layer and the next becomes lost when kT becomes comparable to or larger than the difference in their binding energies to the substrate (see Fig. 9.2). Since the difference in binding of adjacent layers falls off rapidly as the distance increases, it is the higher layer steps that become blurred first, whether because of heterogeneity or of thermal excitation. Thus, the most uniform films are those which display the greatest number of steps, provided the temperature is sufficiently low. By the same token, it is most unusual for the heterogeneity to be so great as to remove any evidence of first layer completion, particularly in the noble gases (where adatom–adatom interactions are generally weaker than adatom–substrate interactions), and it is this remnant that causes the "knee" of the BET vapor pressure isotherm.

Now it is interesting to examine a simple analytic model of a stepwise isotherm and then to use it to illustrate some of the foregoing points. For this we employ the rigid band model isotherm Eq. (9.2.7) which relates the pressure–coverage dependence to the density of states, in the limit of negligible thermal excitation. On an ideal substrate at very low temperature the density of states will be a series of δ functions corresponding to the individual layers. Thus, if the lth layer has binding energy ϵ_l and capacity N_l, its distribution function is $g_l \simeq N_l\,\delta(\mu - \epsilon_l)$. The total distribution function for a film is a sum of such δ functions. The vapor pressure isotherm is then given by

$$(\partial \ln P/\partial N)_T = [kT \sum_l N_l\,\delta(\mu - \epsilon_l)]^{-1}, \qquad (9.7.2)$$

in combination with

$$\mu = kT \ln (P\lambda^3/kT). \qquad (9.7.3)$$

This function yields a staircase curve, with vertical and horizontal sections corresponding to the filling and completion of the individual layers. To make the illustration more explicit, we can adopt a definite progression for the layer energies, such as the familiar inverse cube law. For simplicity the individual layer capacities are assumed equal.

The pressure condition for the filling of the separate layers can then be expressed as a scale of reduced pressures, i.e., relative to the P_0 of the bulk (equivalent to the limiting pressure of a very thick film) as

$$\ln (P_l/P_0) = l^{-3} \ln (P_1/P_0), \qquad (9.7.4)$$

where P_1/P_0 is the partial pressure during the filling of the first layer. At these values of pressure the logarithmic derivative of pressure with respect to N vanishes, and at all other values it diverges. Idealized stepwise isotherms of this sort have been shown by Singleton and Halsey (29). The effects of heterogeneity and of thermal excitation on the isotherm can now be perceived as quite distinct from each other. Heterogeneity enters as additional terms in the distribution function, either as an independent series of δ functions corresponding to other uniform facets coexisting with the primary surfaces, or as a more continuous "background" distribution. Thermal excitation, on the other hand, introduces a rounding of the steps and a tilting of the "treads" due to higher-order terms in the expansion of the Fermi function, Eq. (9.2.4). Of course it is not correct to apply this model to cases in which thermal excitation is large, for then the rigid band approximation breaks down, but it at least serves to illustrate these effects in principle.

In typical adsorption systems there is no evidence of layer formation beyond the "knee" of the isotherm, and beyond this region most isotherms are quite similar in shape. Their similarity has led some people to believe that all effects of heterogeneity are healed out in thicker films, but the following argument shows that this is not true. Suppose that the vapor pressure–thickness relation is given by the Frenkel–Halsey–Hill formula (see Chapter 4)

$$P = P_0(T) \exp (-\alpha/kTd^n), \qquad (9.7.5)$$

where $P_0(T)$ is the vapor pressure of the bulk phase, α is a constant related to the binding energy of the specific gas and substrate, and the exponent $n \simeq 3$. On a nonuniform adsorbate the constant α varies from place to place and yet the vapor pressure P in equilibrium is constant throughout. d must therefore vary so as to keep the exponent constant, which leads immediately to a coupling between the variations of thickness and of binding. Thus, setting the total differential of the exponent to 0, we have

$$\delta d/d \simeq (1/n)(\delta\alpha/\alpha). \tag{9.7.6}$$

The relation is curious and seems at first sight to violate one's physical intuition: it predicts, for example that a film which has a thickness variation of one layer when the mean thickness is five layers will have a variation of ten layers when the mean thickness is fifty layers. That is, rather than the effects of heterogeneity becoming healed out with increasing thickness, they are amplified in proportion to the thickness. As a film gets thicker it becomes "softer" and more responsive to perturbations, whether they be the binding energy variations specifically considered, temperature gradients, or any other influencing field.

Our considerations up to this point have implicitly assumed that the film thickness is small compared to the typical lateral distances of the variations in α. As long as this condition holds the thickness variations continue to grow with thickness, but when d becomes comparable with the scale of lateral heterogeneity of the substrate, an averaging process becomes important; the topmost layers of the film "see" a wider field of view of the surface encompassing a broader range of binding energies. When the film thickness is much larger than the lateral scale of heterogeneity the averaging will become quite effective, and the film thickness tends to a uniform value. In the general case, the substrate heterogeneity cannot be described in terms of a single length but is rather a composite of many Fourier components. Each component induces thickness variations of characteristic wavelength. When the film is much thinner than the wavelength these induced variations increase linearly with thickness; as the thickness exceeds this condition the process reverses and the variations become averaged out. For each component there is a film coverage at which the thickness variations are maximal. The wavelength of maximal variation increases monotonically with film thickness.

REFERENCES

1. I. Langmuir, *J. Amer. Chem. Soc.* **40**, 1361 (1918).
2. S. Z. Roginsky, *C. R. (Dokl.) Acad. Sci. USSR* **45**, 61, 194 (1944).

3. G. Halsey and H. S. Taylor, *J. Chem. Phys.* **15**, 624 (1947).

4. R. Sips, *J. Chem. Phys.* **16**, 490 (1948).

5. M. J. Sparnaay, *Surface Sci.* **9**, 100 (1968).

6. D. N. Misra, *Surface Sci.* **18**, 367 (1969).

7. G. F. Cerofolini, *J. Low Temp. Phys.* **6**, 473 (1972).

8. J. Appel, *Surface Sci.* **39**, 237 (1973).

9. S. Ross and J. P. Olivier, "On Physical Adsorption." Wiley (Interscience), New York, 1964.

10. T. L. Hill, *J. Chem. Phys.* **17**, 762 (1949).

11. N. N. Roy and G. D. Halsey, Jr., *J. Low Temp. Phys.* **4**, 231 (1971).

12. G. A. Stewart and J. G. Dash, *Phys. Rev.* **A2**, 918 (1970).

13. C. E. Campbell, J. G. Dash, and M. Schick, *Phys. Rev. Lett.* **26**, 966 (1971).

14. A. Widom and J. B. Sokoloff, *Phys. Rev.* **A5**, 475 (1972).

15. A. D. Novaco, *J. Low Temp. Phys.* **9**, 457 (1972).

16. Y. Imry, D. J. Bergman, and L. Gunther, in *Proc. Int. Conf. Low Temp. Phys., 13th* (R. H. Kropschot and K. D. Timmerhaus, eds.). Plenum, New York 1974.

17. M. M. Dubinin and L. V. Radushkevich, *C. R. (Dokl.) Acad. Sci. USSR* **55**, 327 (1947).

18. M. G. Kaganer, *Dokl. Akad. Nauk USSR* **116**, 251 (1957).

19. H. G. Kaganer, *Proc. Acad. Sci. USSR* **122**, 663 (1958).

20. N. Hansen, *Vakuum-Tech.* **3**, 70 (1962).

21. J. P. Hobson and R. A. Armstrong, *J. Phys. Chem.* **67**, 2000 (1963).

22. F. Ricca and R. Medana, *Ric. Sci.* **4**, 617 (1964); F. Ricca, R. Medana, and A. Bellardo, *Z. Phys. Chem. NF* **52**, 276 (1967).

23. J. P. Hobson, in "The Solid-Gas Interface" (E. A. Flood, ed.), Chapter 14. Dekker, New York, 1967.

24. B. A. Gottwald, in "Adsorption-Desorption Phenomena" (F. Ricca, ed.), pp. 19–32. Academic Press, New York, 1972.

25. J. P. Hobson and R. Chapman, in "Adsorption-Desorption Phenomena" (F. Ricca, ed.), pp. 33–48. Academic Press, New York, 1972.

26. C. V. Heer, *J. Chem. Phys.* **55**, 4066 (1971).

27. S. Brunauer, P. H. Emmett, and E. Teller, *J. Amer. Chem. Soc.* **60**, 309 (1938).

28. D. M. Young and A. D. Crowell, "Physical Adsorption of Gases." Butterworths, London and Washington, D.C. 1962.

29. J. H. Singleton and G. D. Halsey, Jr., *J. Phys. Chem.* **58**, 330, 1011 (1954); *Can. J. Chem.* **33**, 184 (1955); C. F. Prenzlow and G. D. Halsey, Jr., *J. Phys. Chem.* **61**, 1158 (1957).

30. A. Thomy and X. Duval, *J. Chim. Phys. Physicochim. Biol.* **66**, 1966 (1969); **67**, 286, 1101 (1970).

31. Y. Larher, *J. Phys. Chem.* **72**, 1847 (1968); *J. Colloid Interface Sci.* **37**, 836 (1971).

32. R. H. Fowler, "Statistical Mechanics." Cambridge Univ. Press, London and New York, 1936.

33. L. D. Landau and E. M. Lifshitz, "Statistical Physics." Pergamon, Oxford, 1958.

34. J. G. Dash, *Phys. Rev.* **A1**, 7 (1970).

35. L. E. Drain and J. A. Morrison, *Trans. Faraday Soc.* **48**, 316 (1952).

36. H. Freundlich, *Trans. Faraday Soc.* **28**, 195 (1932); G. D. Halsey, Jr., in "The Solid-Gas Interface" (E. A. Flood, ed.), Vol. 1, Chapter 16. Dekker, New York, 1967.

37. M. Polanyi, *Trans. Faraday Soc.* **28**, 316 (1932).

38. T. L. Hill, *Advan. Catal.* **4**, 211 (1952).

39. C. Kittel, "Elementary Statistical Physics," p. 95. Wiley, 1958.

40. S. S. Shinozaki and A. Arrott, *Phys. Rev.* **152**, 611 (1966).

41. D. W. Princehouse, *J. Low Temp. Phys.* **8**, 287 (1972).

42. J. E. Robinson, *Phys. Rev.* **83**, 678 (1951).

43. F. London, "Superfluids," Vol. 2. Wiley, 1954 (Reprint, Dover, New York, 1964, Appendix).

44. L. Goldstein, *J. Chem. Phys.* **9**, 273 (1941).

45. W. E. Lamb and A. Nordsieck, *Phys. Rev.* **59**, 677 (1941).

46. A. Widom, *Phys. Rev.* **168**, 150 (1968); **176**, 254 (1968).

47. J. J. Rehr and N. D. Mermin, *Phys. Rev.* **B1**, 3166 (1970).

48. J. M. Thomas, E. L. Evans, and J. O. Williams, *Proc. Roy. Soc. London A* **331**, 417 (1972).

49. D. L. Goodstein, W. D. McCormick, and J. G. Dash, *Phys. Rev. Lett.* **15**, 447 (1965).

50. W. D. McCormick, D. L. Goodstein, and J. G. Dash, *Phys. Rev.* **168**, 249 (1968).

51. P. Mahadev, M. F. Panczyk, R. A. Scribner, and J. G. Daunt, *Phys. Lett.* **A41**, 221 (1972).

52. J. G. Daunt, *Phys. Lett.* **A41**, 223 (1972).

53. D. F. Brewer, *J. Low Temp. Phys.* **3**, 205 (1970).

54. D. F. Brewer, A. Evenson, and A. L. Thomson, *J. Low Temp. Phys.* **3**, 603 (1970).

55. M. Bretz, J. G. Dash, D. C. Hickernell, E. O. McLean, and O. E. Vilches, *Phys. Rev.* **A8**, 1589 (1973); **A9**, 2814 (1974).

56. R. L. Elgin and D. L. Goodstein, *Phys. Rev.* **A9**, 2657 (1974).

57. G. D. Halsey, *Discuss. Faraday Soc.* **8**, 54 (1950).

58. G. D. Halsey, *Advan. Catal.* **4**, 259 (1952).

10. Superfluidity

This topic occupies a place of special importance in adsorption owing to the intense theoretical interest and experimental activity in thin film superfluidity, reaching back to the 1930s and continuing to the present day (1). But in spite of its long history only some qualitative features of the problem are unequivocally established, while others are in dispute. For example, it had been widely accepted for many years that the superfluidity of bulk liquid ^{4}He is progressively modified by confinement in channels or films of decreasing thickness, until at thicknesses of a few atomic layers all traces of superfluidity disappear. These experimental results were in accord with the qualitative predictions of several different theoretical models, that superfluidity is suppressed and finally eliminated as 2D geometry is approached. However, recent more delicate experiments at very low temperatures now suggest that traces of superfluidity persist in films of one or two atomic layers at very low temperatures. New theoretical developments have also made a break with the past, providing arguments for the persistence of superfluidity even in the strictly 2D limit. In this chapter we give a historical review of both theory and experiment, leading up to the current state of the field.

10.1 THEORY

10.1.1 Bose Gas Models

The ideal Bose gas was the earliest model for the lambda transition in bulk ^{4}He, and it has been the basis for a number of calculations on thin He films. The first of these studies was made by Osborne (2), who treated the case of slab geometry $L \times L \times D$, $D \ll L$. He analyzed the 2D limit in some detail, and showed that although there could be no Bose–Einstein condensation in the usual sense, for a finite L system there is a moderately abrupt accumulation of particles into the ground state at sufficiently low temperature, the accumulation temperature varying as $(\ln N)^{-1}$. For films of finite thickness, Osborne argued that the ordinary criterion of BE condensation breaks down only when the film becomes so thin that the zero point energy becomes comparable to kT: thus when the energy separation between levels corresponding to higher excitations in the direction normal to the plane of the film becomes greater than kT, the film is essentially 2D. For parameters corresponding to liquid ^{4}He, Osborne's criterion is reached at $\sim$10 Å, which is comparable to the thickness of ^{4}He films at which superfluid phenomena are strongly affected or suppressed.

Ziman (3) reexamined the slab model, improving over Osborne's integral approximations for the summations over the discrete single-particle levels. His result was sharply different from Osborne's in an important respect: the condensation temperature of a layer of finite thickness, *however large*, and fixed density varies as $(\ln N)^{-1}$ as $L \to \infty$. Thus, a film of finite thickness and of infinite lateral extent is 2D, and its condensation temperature $T_0 = 0$. Ziman concluded that, if the ideal gas has any validity as a model for real ^{4}He films, the experimental observations of superfluidity in thin films of large area imply that there must be some finite upper limit on the effective size of the cooperating regions. If the "maximal assemblies" have empirical dimensions $\sim$7 $\times$ 10^{-6} cm, the slab model yields accumulation temperatures which are in reasonable agreement with experiment.

Mills (4), Khorana and Douglass (5), and Goble and Trainor (6) examined the ideal Bose gas in various box geometries, evaluating the properties by computer. Mills considered rectangular prisms

$D \times D \times L$, with 10 Å $\lesssim D \lesssim 100$ Å and 200 Å $\lesssim L \lesssim 5000$ A, and particle density equal to liquid ^{4}He. He found that for $D = 100$ Å, $L = 1000$ Å, the ground state occupation n_0 had a temperature dependence similar to the 3D ideal gas. When D is decreased to 10 Å, n_0 suffers a rapid depletion as T is increased above 0, but n_0 remains small and finite to much higher temperatures. Mills pointed out that the behavior for small D approaches the characteristics of a 1D system, and that this correspondence occurs when the energy separation between the lowest levels of excitations in the D direction becomes comparable with $kT_0(3D)$. Mills argued that the evolution of 1D characteristics can be gauged from the densities of states corresponding to the D-wise and L-wise excitations: if $\rho_2(E)$ is the 2D density of states of the square cross section and $\rho_1(E)$ is the 1D density of states along L, large deviations from 3D behavior should occur when

$$\rho_2(kT)/\rho_1(kT) = (D^2/hL)(2mkT) \ll 1. \qquad (10.1.1)$$

This condition is consistent with Ziman's conclusion that the characteristics of the lower dimensional system "win out" when the larger dimension(s) is extended to infinity.

Goble and Trainor (6) studied slab geometries for several D values at $L = 750$ Å and particle density equal to liquid ^{4}He. In addition to n_0 they evaluated the first excited state occupation n_1, the chemical potential and the heat capacity. They found that the temperature dependence of n_0 in all of the finite assemblies was similar to Mills' result: a rapid change at low T and a slowly varying tail extending to temperatures above the T_0 of the infinite 3D gas. The specific heat was particularly interesting: broad maxima were found for $D \lesssim 10$ Å, becoming narrower with increasing D and forming a progression trending toward the sharp cusp of the infinite 3D gas. For the finite systems, however, none of the computed quantities displayed discontinuities in value or temperature derivative at any T. Goble and Trainor examined six alternative definitions for T_0 that might provide relatively sharp criteria for transitions in finite systems. All of the definitions yielded different values of T_0 except in the infinite 3D limit. Moreover, although five criteria gave T_0 falling with decreasing D in qualitative agreement with experiment, the specific heat maxima shifted to *higher* T. Goble and Trainor concluded that Ziman's idea of maximal assem-

blies has little promise for explaining the behavior of He films. Nevertheless they proposed that Ziman's model might have some merit in a
more general sense, for they observed that the absolute heights of the
heat capacity maxima first increased as D was reduced and then gradually decreased to zero in the $D = 0$ limit. They took this reversal as
indication of an evolution from essentially 3D to 2D behavior, with
the value $D \sim 70$ Å at the reversal as a "statistical correlation length"
in the film. In subsequent studies (7) they explored the effects of hard-
core interactions in a low density approximation, and they found similar D dependences as in the noninteracting gas.

Dewar and Frankel (8) helped to explore some of the above effects
by carrying out calculations for fermions similar to those that Goble
and Trainor undertook for bosons. They computed heat capacities for
noninteracting particles in $L \times L \times D$, $D \ll L$ geometry, and showed
that the behavior of the system depends upon three characteristic
lengths: D, the interparticle spacing, and the thermal de Broglie wavelength. At sufficiently low densities, the specific heat of the system at
decreasing T changes from effectively 3D to 2D classical behavior and
then at still lower T, to a 2D quantum gas: this is a result of the discrete momenta along D, which freeze out before the onset of quantum
degeneracy. At high densities, the effects of dimensionality and degeneracy are present simultaneously, and the decrease of the specific
heat in the quantum regime is influenced by the finite D. A particularly interesting result was seen at low density, where the finite D
caused a maximum in the heat capacity above the classical value
$\frac{3}{2}k$. The maximum arises from the finite energy level separation of
the D-wise single-particle modes, analogous to the Schottky anomaly
of systems with low-lying discrete states.

10.1.2 Long-Range Order in Finite Geometries

Hohenberg (9) proved that long-range order of the type conventionally associated with superfluidity (10–13) cannot take place in a
2D Bose *liquid* at finite temperature. Basing his proof on a rigorous
inequality due to Bogoliubov (13), Hohenberg showed that in lower
dimensional systems the fluctuations of the order parameter diverge
at $T > 0$, destroying the long-range order of the average momentum.

The distinction between 2D and bulk systems in this regard lies in the abundance of low energy excitations in 2D, which is directly due to the differences in the momentum dependence of the phase space. It is this same effect of changed dimensionality that is responsible for the absence of long-range crystalline order in 2D systems (see Chapter 7), and it is the same theorem of Bogoliubov which was applied by Mermin to the crystalline case. The theory is applicable to liquids, i.e., interacting assemblies, of quite general character. There are only three assumptions involved in the proof: that superfluid long-range order is equivalent to nonvanishing anomalous averages $\langle \psi \rangle$ or $\langle \psi \psi \rangle$ of the wavefunction; that the system is infinite in lateral extent, and that there are no fields or other conditions destroying translational invariance. It is the last condition which explains why the Hohenberg proof does not exclude a pseudo condensation caused by rotation or lateral fields (see Chapter 9, Section 9.5).

Krueger (14) and Chester et al. (15) demonstrated that condensation cannot occur in *finite* thickness films if their lateral dimensions are infinite. Their proofs are based upon the same type of argument which Hohenberg applied to strictly 2D systems. But Krueger noted that laboratory films are poor approximations to "partially finite" geometries, since the theoretical divergence varies only logarithmically on lateral size. For example, in a film of 45 Å thickness and 1 cm lateral dimension there would be appreciable ground state occupation at $T = 1°K$. Imry (16) reached similar conclusions concerning the effect of finite dimensions, and noted that the transition from effectively 3D to 2D behavior sets in at thicknesses of order $\ln N$. Chester, Fisher, and Mermin showed that the absence of condensation in partially finite geometry is not dependent on special boundary conditions or confining potentials. Their suggestion for a closer correspondence with experiment is that there might be a more subtle kind of "weak long-range order" which could persist as $L \to \infty$, such that although the ground state density vanishes in that limit the integral occupation over the entire film diverges. This divergence corresponds to an infinite "off-diagonal susceptibility" which has been suggested (17) in the analogous case of a 2D Heisenberg ferromagnet. Chester et al. also remarked that their proof presumed a homogeneous system; Hohenberg noted the same restriction in his treatment of 2D geometry.

Jasnow and Fisher (18) reinspected the proofs of the absence of long-range order in 1D and 2D systems of Bose particles, with particular attention to the following questions:

(a) How does the static order–order correlation function $\sigma(\mathbf{r}, \mathbf{r}')$ behave for large spatial separations of the arguments?

(b) Can one avoid the usual mathematical technique of introducing a symmetry-breaking field (such as a magnetic field in the case of spin systems) and then proceeding to the thermodynamic limit (volume $V \rightarrow \infty$), involved in the preexisting proofs of absence of long-range order in 2D systems?

Their analysis assumed slab geometry, with hard-wall boundaries at the planes and no fields normal to the planes. They concluded

(a) that the integral of the normal-averaged correlation function, if monotonically decreasing with increasing r, vanishes at least as fast as $(\ln r)^{-1}$, and

(b) the use of the symmetry-breaking field may be avoided altogether.

Applying their results to the He film problem, Jasnow and Fisher found that the proof could not rule out "effective" Bose condensation in laboratory-sized films thicker than ~ 1.5 Å.

10.1.3 Theory of Ginzburg and Pitaevskii, and Gross

Ginzburg and Pitaevskii (19) took a different approach to the He film problem than the preceding papers. Their work is based on Landau's phenomenological theory of second-order phase transitions (20). In this theory one expands the Gibbs free energy in a series of powers of some order parameter ψ which in equilibrium is zero on one side of the transition. Its nonzero value on the other side of the transition is determined by the condition that the free energy is a minimum with respect to the order parameter. Ginzburg and Pitaevskii assumed that in liquid helium the order parameter is simply related to the superfluid density ρ_s; specifically, they proposed a complex function

$$\psi(\mathbf{r}) = \eta e^{i\phi}, \qquad (10.1.2)$$

with

$$\rho_s = m|\psi|^2, \qquad \mathbf{v}_s = (\hbar/m)\,\nabla\phi, \qquad (10.1.3)$$

where m is the mass of the He atom mass and $\mathbf{v}_s$ is the velocity of superfluid. They then followed the standard lines of Landau's theory, including the assumption that the expansion coefficients can themselves be expanded in a power series in $(T - T_\lambda)$ near the transition temperature. They evaluated the first two parameters by comparison with measured properties of ^{4}He. The numerical values of the coefficients indicate that the order parameter has a "healing length" $l_{cm} \simeq 4 \times 10^{-8}(T_\lambda - T)^{-1/2}$. Therefore, in order that the macroscopic theory be applicable, l must be much greater than interatomic distances and hence the temperature must be quite close to T_λ. The He film problem was treated by assuming that ψ vanishes at both the solid boundary and the free surface, and that the film is uniform over its lateral extent. They derived the result for the downward shift ΔT_λ of the transition temperature in a film of thickness d:

$$\Delta T_\lambda \approx 2 \times 10^{-14}(d_{cm})^{-2} \quad {}^\circ\text{K}.$$

Noting that the result can only be applied to films of thickness $d \gg$ (the interatomic spacings in liquid He), GP found moderate agreement with the experimental onset temperatures for relatively thick films.

Josephson (21) pointed out that the order parameter in liquid He may be more properly related to the density of the ground state rather than to the density ρ_s. With this revision, Josephson showed that the experimentally observed logarithmic singularity of the bulk specific heat was consistent with the measured temperature dependence of ρ_s near T_λ; experimentally, $\rho_s \propto (T - T)^{2/3}$, whereas the GP theory predicts $\rho_s \propto (T_\lambda - T)$.

Mamaladze (22) incorporated Josephson's modification in a revision of the GP theory of onset in thin films: his formulas for the healing length l and lambda point shift ΔT_λ in films are

$$l = 2.73 \times 10^{-8}(T_\lambda - T)^{-2/3} \quad \text{cm}, \qquad (10.1.4)$$

$$\Delta T_\lambda = 2.5 \times 10^{-11}(d_{cm})^{-3/2} \quad {}^\circ\text{K}. \qquad (10.1.5)$$

The theory predicts shifts within about a factor of two of the measured onset temperatures. In addition, Mamaladze concluded that the predicted $d^{-3/2}$ law is in better agreement with the shift of the specific heat maximum than the d^{-2} law of the original GP theory.

Gross independently arrived at essentially the same theory as Ginzburg and Pitaevskii, beginning first from the weakly interacting Bose gas (23), and then from a model of a quantum fluid (24). He applied the theory to several specific problems; rigid parallel walls, a rotating cylinder, vortex line solutions, and the motion of an ion through the superfluid.

One notices that dimensionality apparently plays no role in the Ginzburg, Pitaevskii–Gross theory. Effects of lower dimensional disorder of the type considered in Section 10.1.2 might enter if one took account of low-lying excitations. However, if they were treated in the manner discussed in the next section, the existence of superfluidity at finite temperatures would not necessarily be destroyed.

10.1.4 Long-Range Coherence and Topological Order

The ideas of Lasher (25), Berezinskii (26), and Kosterlitz and Thouless (27) contrast sharply with preceding theories in predicting a finite superfluid transition temperature for a 2D Bose liquid, even in the infinite size limit and in the absence of lateral fields. The theory depends upon the properties of a kind of order that Lasher terms "long-range coherence" and Kosterlitz and Thouless call "topological order," which is relevant to crystalline and magnetic as well as superfluid order in lower dimensional systems. Its application to the melting problem in films is discussed in Chapter 7. Kosterlitz and Thouless present the argument with respect to superfluidity as follows.

Assume that a 2D superfluid exists at $T = 0$. As the temperature is raised, thermal fluctuations are excited, causing the phase of the order parameter or condensate wavefunction to vary with position, so that the type of order defined by Penrose and Onsager (10, 11) does not occur. But if a condensate wavefunction can be defined locally, the variation of its phase with position can be explored from one region to another, and as long as the phase remains correlated between the regions it indicates a continuity of wavefunction that can be associated with superfluidity. The thermal excitations that destroy superfluidity are quantized vortices (28) of circulation

$$\oint \mathbf{v} \cdot d\boldsymbol{l} = h/m, \tag{10.1.6}$$

where m is the effective mass of an atom in the fluid. There is a change in phase of 2π for each circuit around the vortex. The energy of a vortex increases logarithmically with the size of the system. For a film of superfluid areal density n_S the kinetic energy is

$$E = \int_{r_0}^{R} \tfrac{1}{2} n_S m v^2 \, d^2 r,$$

and substituting $v(r) = \hbar/mr$ from the circulation quantization formula, we obtain

$$E = (\pi n_S \hbar^2/m) \ln (R/r_0). \tag{10.1.7}$$

Here r_0 represents the radius of the core of the vortex, estimated to be on the order of an atomic spacing for liquid ^{4}He and R is the size of the system. The condition for the creation of an excitation is that the free energy of the system should be a minimum. The entropy associated with the vortex depends logarithmically on the size of the system, for since there are approximately R^2/r_0^2 positions at which it may be placed, the entropy is

$$S \simeq 2 k_B \ln (R/r_0). \tag{10.1.8}$$

Thus, both the energy and entropy vary as $\ln R$, so that the energy term dominates at low T and no vortices will be created, whatever the size of the system. There is a definite temperature at which the entropy term takes over, and this is approximately given by

$$k_B T_c \simeq \pi n_s \hbar^2/2m. \tag{10.1.9}$$

For $m = m(^4$He$)$ and n_s equal to the density of a completed monolayer, $T_c \simeq 2°$K. But n_s would be reduced by other thermal excitations such as phonons, so that T_c would be lower. It is not clear how the proper value of n_s is to be estimated nor how interactions between vortices will affect the order of the transition. By a mean field approximation on a model system Kosterlitz and Thouless obtained a very weak singularity in the free energy but none in the heat capacity at the transition. Taking account of the typical errors caused by mean

field approximations, it is believed that the actual heat capacity may
have a weak singularity.

10.2 EXPERIMENTS

10.2.1 Heat Capacity

One of the most striking manifestations of the superfluid transition
in liquid ^{4}He is the λ-point singularity in the heat capacity at $T_\lambda =$
2.17°K (1). It marks the temperature of onset of all anomalous transport coefficients in bulk liquid, but the demarcation is less distinctive
in films, and it appears to vary with experimental conditions. The
pioneering measurement of the heat capacity of thin ^{4}He films was
made by Frederikse in 1949 (29, 30). The films were adsorbed on a fine
powder of Fe_2O_3 (jeweler's rouge), and several coverages were studied
in the range from about three to twelve atomic layers. Frederikse's
results, reprinted in Fig. 10.1, show that the λ-point anomaly is progressively rounded and shifted to lower temperature with decreasing
thickness. At the lowest coverages studied there is no trace of a peak
remaining, and the heat capacity increases monotonically with T, approximately as T^2. Frederikse's results were the primary stimulus for
early theoretical interest in the effects of dimensionality and finite
geometry on Bose condensation, and for experimental studies of transport in unsaturated ^{4}He films.

Specific heat curves similar to Frederikse's were obtained by Mastrangelo and Aston using TiO_2 powder (31), by Brewer et al. using
porous Vycor glass (32), and by Symonds on Vycor (33, 34).

Bretz obtained qualitatively different results with exfoliated graphite (Grafoil) substrate (35). These results, seen in Fig. 10.2, show distinctive sharp "corners" at both T_λ and at lower temperatures, for
thicknesses in the range ~5–11 layers. The peaks are not rounded as
in the earlier work on other substrates, resembling instead the bulk
liquid anomaly with the top of the peak snipped off, the truncation
being more severe in thinner films. The bulk liquid T_λ seems to retain some significance down to five or so layers, with very little shift
at higher thicknesses. At temperatures below the lower corners the
thermal response time of the calorimeter was significantly shorter,

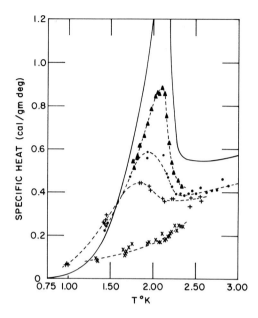

FIG. 10.1 Specific heat of ⁴He films adsorbed on jeweler's rouge, measured by
Frederikse (29, 30). Films adsorbed in porous Vycor glass have similar shapes
(32–34). Coverages in statistical (BET) layers are: (×) 3–4 layers; (+) 5–6 layers;
(●) 7–9 layers; (▲) 9–12 layers. The solid line corresponds to bulk liquid.

suggesting a marked increase in thermal conductivity of the film
associated with superfluidity.

The contrast between Bretz' results and previous work appears due
to the greater uniformity of the graphite substrate. The rounded and
shifted anomalies of Fig. 10.1 can be produced by a distribution of
truncated peaks, the distribution being caused by variations in thickness due to substrate heterogeneity.

Notwithstanding the importance of and interest in Bretz' results,
they have not been explained. Doniach (36) and Penrose (37) have
proposed explanations for the dependence of the lower "transition"
temperature on thickness; Doniach by an argument based on Ginzburg–Landau model field theory, and Penrose in terms of finite geometry effects on Bose liquid condensation. The distinctive shapes of the
peaks are not understood nor what is the nature of the "intermediate
state" between the lower transition temperature and T_λ.

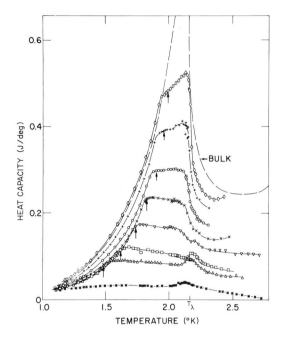

FIG. 10.2 Specific heat of ^{4}He films adsorbed on exfoliated graphite, measured by Bretz (35). n = ($\diamondsuit$) 11.20; ($+$) 9.70; ($\bigcirc$) 8.64; ($\times$) 7.80; ($\triangledown$) 6.77; ($\square$) 5.69; ($\triangle$) 5.25; ($*$) 3.62. n is the average thickness in layers.

10.2.2 Thermal and Mass Transport

In relatively thick ($\sim$100 layer) films which form on walls in contact with liquid ^{4}He, anomalous thermal and mass flow sets in at the λ point. At $T < T_\lambda$ the film can be made to flow with negligible resistance up to a critical velocity of a few centimeters a second. The portion of the film that flows is the superfluid fraction which carries no entropy. In a thermal gradient the superfluid tends to move toward the higher temperature, where a portion is converted to normal fluid. The film that evaporates at the higher temperature can flow in the gas phase to the colder region where it condenses. The overall convective process is an effective mechanism for heat transport, and is useful as a detector of superfluid onset.

The onset temperature T_0 of superfluidity is depressed below T_λ as film thickness is reduced significantly below the saturated value. This

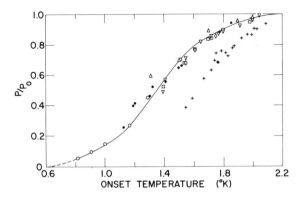

FIG. 10.3 Superfluid onset temperatures of ⁴He films for several different substrates and experimental methods, plotted against partial pressure. (●, ○, △, □, ▽, ◇) various mass transport, thermal transport, persistent current, and third sound experiments on glass and metal substrates (38–43): (+) mass transport on graphite (45).

is a common result of the many studies that have been made using various techniques and substrates (38–49). A collection of results is illustrated in Fig. 10.3. Quantitative agreement is poor, however, both with regard to different methods of detection and in comparisons between experiment and theory. The difficulties are probably due in part to substrate heterogeneity which must be present in all of the experiments. It is noteworthy that the results on exfoliated graphite (45) are noticeably different from the rest of the data, a difference which can be attributed to the greater uniformity of the graphite surface. Two other aspects of the graphite results are distinctive: a gradual "precursor" region above $T = 2.2°K$ in which the flow rate grows gradually with falling T above the trend of low coverage films, and a sudden increase occurring at or near T_λ in films as thin as ∼3 layers. The change in flow rate that occurs at the λ temperature of bulk liquid is entirely distinct from the more dramatic increases at lower temperatures. The occurrence of two separate "transitions," one at T_λ and the other at a lower temperature which increases monotonically with film thickness, resembles Bretz' heat capacity results on the same type of substrate. Long and Meyer (38, 39) also observed a change in flow at T_λ in thin film mass transport on glass and metal surfaces, but were not able to observe this behavior in more than one type of experimental arrangement.

10.2.3 Third Sound

A unique form of coupled thermal and density excitation known as third sound (50, 51) exists in superfluid films. It is a surface wave involving superfluid motion caused by thermal gradients while the normal component of the film remains stationary due to viscosity. The excess superfluid moving toward the warmer regions increases the film thickness locally, and this increase is opposed primarily by the van der Waals attraction to the substrate. Therefore the principal factors in the wave equation are the superfluid inertia, the thermomechanical pressure, and the substrate attraction. With the assumption of the Frenkel–Halsey–Hill formula relating thickness to partial pressure P/P_0, the equation for third sound velocity u_3 is

$$u_3{}^2 = 3(\bar{\rho}_s/\rho_{\mathrm{surf}})(k_B T/m) \ln (P_0/P), \qquad (10.2.1)$$

where $\bar{\rho}_s$ is the average superfluid density in the film and ρ_{surf} is the total density at the surface of the film (estimated to be within a few percent of the bulk density). The first observation of third sound in a thick film was by Everitt et al. (52) using optical methods of observation. Later developments by Rudnick and collaborators (43, 48, 53) utilized superconducting transducers for delicate measurements on very thin films, and the results have stimulated considerable theoretical interest. Rudnick's studies are usually made by pulse methods, which permit both velocity and attenuation to be measured. Ratnam and Mochel (54) have experimented with a resonant cavity technique which is extremely sensitive to velocity changes but is somewhat more difficult to interpret. Both groups of investigators have typically used substrates of Pyrex glass.

The third sound results agree with mass and thermal transport data in the general depression of T_0 with decreasing film thickness of which sample values are included in Fig. 10.3. A particular advantage of the third sound technique is the theoretical connection with the superfluid density in the film via Eq. (10.2.1). The experimental results are surprising in that it appears that $\bar{\rho}_s/\rho_{\mathrm{surf}}$ is not zero at onset but is a quite appreciable fraction ($\sim$0.4) when signals first become apparent. That is, the signals are observed to attenuate as the thickness is reduced below the onset value, while the velocity u_3 remains at a finite value. The large attenuation at onset cannot be explained in terms of the known thermal and hydrodynamic mechanisms existing in liquid He(II) (55, 56). If the superfluid fraction changed discontinuously at onset there would be an anomaly in the vapor pressure at T_0, but in an experiment designed to measure the jump it could not be detected (57). It has been proposed that the anomalous attenuation is produced by heterogeneities (58). Discussion of this explanation is given in Section 10.2.5.

10.2.4 Persistent Currents

The most sensitive experimental detection of superfluidity is by observation of persistent circulating currents, analogous to persistent electric currents of superconducting loops. Although the circulating state is only metastable (59), the decay of macroscopic motion is so slow that it is experimentally observable as an essentially permanent angular momentum (60–63). The method of detection developed by Reppy and his collaborators (63, 64) is based on the gyroscopic torque produced by a ring of circulating superfluid when it is slightly changed in orientation. This device, originally developed for studies on bulk liquid He(II), has been applied to studies of superfluid onset in thin films (42, 49). The onset temperature–thickness dependence on Vycor substrates was found to be similar to results of other experiments, although quantitative disagreements amounted to as much as 30% in onset thickness. An especially interesting finding in the gyroscope experiments is that a persistent current appears quite abruptly with increasing film thickness. When the coverage is increased beyond some critical thickness while temperature is held constant, the gyroscopic torque jumps from zero to a finite value. The suddenly appearing persistent angular momentum corresponds to an increase in superfluid mass greater than the experimental increment in the total mass of the film. This behavior, which could be related to the anomalous attenuation of third sound at onset, can also be qualitatively understood in terms of substrate heterogeneities, as discussed in the next section.

10.2.5 Heterogeneity and Superfluid Connectivity

A model explaining the role of substrate heterogeneity in superfluid
onset experiments was proposed by Dash and Herb (58) and explored
in greater detail by Cole *et al.* (65). Originally stimulated by a study
of onset using a high frequency shear mode quartz microbalance (44),
the model assumes that all of the experimental substrates are heterogeneous to some degree. Therefore an adsorbed film is uneven in thickness, greater thickness being associated with regions of greater binding
(see Chapter 9, Section 9.7). We may assume from many experiments
that there is some intrinsic relation between the film thickness at onset, which might have a trend similar to Fig. 10.3. Therefore, in a real
(heterogeneous) film at onset there is a distribution of film thickness
and consequently of superfluidity. The particular distribution depends
on the substrate and the method of detection. In techniques that are
especially sensitive to attenuation such as persistent currents the
practical onset conditions require a substantial connectivity of superfluidity throughout the ring. In other methods involving continually
applied driving forces the connectivity required is much weaker. But
in each type of experiment just before onset there is actually some
superfluidity already present, although isolated by boundary regions
of normal (i.e., nonsuperfluid) film. As the coverage is increased beyond some point the normal regions are made thick enough to become
superfluid. This point may be reached over a very narrow vapor pressure interval by a sufficient fraction of the normal regions for the
superfluid connectivity to increase quite suddenly to detectable limits.
If this happens onset will be abrupt in the manner seen in the persistent current experiments. The changing connectivity at onset can
also appear as a kind of attenuation for third sound signals.

The connectivity model implies a frequency dependence for the observation of superfluid onset by the microbalance technique. This results from the fact that the vibrating crystal in shear vibration is
sensitive (i.e., inertially loaded) to normal fluid locked to the surface
by its viscosity and also to superfluid regions much smaller than a
wavelength of third sound. Since $\lambda_3 = u_3/f$, where f is the frequency
of the crystal, it was predicted that onset should appear "earlier"
(i.e., at smaller thicknesses and/or higher temperatures) at higher fre-

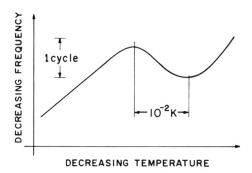

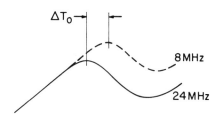

FIG. 10.4 Frequency response of a quartz crystal microbalance near the onset temperatures of ⁴He films adsorbed on its surface (66). The local maxima in the curves are identified with superfluid onset. Measurements at the fundamental mode and third harmonic show that apparent onset is at higher temperature (or lower thickness) for higher crystal frequencies. At higher frequencies the technique is sensitive to smaller (lateral) regions of superfluid film. The frequency dependence of onset was predicted on the basis of a model of substrate heterogeneity and superfluid connectivity (58).

quencies. An experimental test of the prediction has confirmed the effect (66) (see Fig. 10.4).

10.3 SUMMARY AND PROSPECT

It should now be evident to the reader that, as noted in the beginning of this chapter, thin-film superfluidity is still poorly understood in spite of many years of theoretical and experimental activity. A wide variety of novel and delicate techniques has been applied to the detection of onset and the measurement of transport properties of superfluid films. Theoretical models have a great deal of diversity, and the calculations have become extremely sophisticated. The lack of correspondence between theory and experiment seems not due to any lack of ingenuity or expertise. On the contrary, we believe that the major reason for the persistence of the puzzle is relatively simple; the common neglect of heterogeneity as an important perturbing influence in all real films. In earlier chapters there are many examples of how heterogeneity is a feature of all real surfaces, and in this chapter we summarize arguments put forth to explain how heterogeneity can affect thin-film superfluidity. Predictions based on the model have been put to experimental test, and although the tests are not completely conclusive, they bear out the predictions, giving strong encouragement for further experiments and elaborations of the model.

Assuming that heterogeneity is indeed as important to superfluidity in real films as it now seems to be, this does not imply that the topic is any less fundamental or interesting. On the contrary, the subject can then develop new and more challenging aspects for both experimentalists and theorists, and although it will be more demanding it should bring greater rewards. It should help resolve the long-standing questions of superfluidity in a single layer and the evolution of bulk properties in progressively thicker films. Heterogeneous superfluid films might be useful in the more general study of disordered systems, which are becoming increasingly important in their fundamental and practical aspects. And finally, perhaps thin-film superfluidity can itself be developed into a probe of surface character, being particularly sensitive to subtle variations in substrate properties.

REFERENCES

1. F. London, "Superfluids," Vol. II. Wiley, New York, 1954 (Reprint Dover, New York, 1964); K. R. Atkins, "Liquid Helium," Cambridge Univ. Press, London and New York, 1959; J. Wilks, "Liquid and Solid Helium." Oxford Univ. Press (Clarendon), London and New York, 1967; R. J. Donnelly, "Experimental Superfluidity." Univ. of Chicago Press, Chicago, Illinois, 1967; W. E. Keller, "Helium-3 and Helium-4," Plenum Press, New York, 1969.

2. M. F. M. Osborne, *Phys. Rev.* **76**, 396 (1949).

3. J. M. Ziman, *Phil. Mag.* **44**, 548 (1953).

4. D. L. Mills, *Phys. Rev.* **134**, A306 (1964).

5. B. M. Khorana and D. H. Douglass, *Phys. Rev.* **138**, A35 (1965).

6. D. F. Goble and L. E. H. Trainor, *Phys. Lett.* **18**, 122 (1965); *Can. J. Phys.* **44**, 27 (1966); *Phys. Rev.* **157**, 167 (1967).

7. D. F. Goble and L. E. H. Trainor, *Can. J. Phys.* **46**, 839, 1867 (1968).

8. R. L. Dewar and N. E. Frankel, *Phys. Rev.* **165**, 283 (1968).

9. P. C. Hohenberg, *Phys. Rev.* **158**, 383 (1967).

10. O. Penrose, *Phil. Mag.* **42**, 1373 (1951).

11. O. Penrose and L. Onsager, *Phys. Rev.* **104**, 576 (1956).

12. C. N. Yang, *Rev. Mod. Phys.* **34**, 694 (1962).

13. N. N. Bogoliubov, Dubna rep. 1962 [*German transl.: Phys. Abhandl. S.U.* **6**, 1, 113, 229 (1962)].

14. D. A. Krueger, *Phys. Rev. Lett.* **19**, 563 (1967); *Phys. Rev.* **172**, 211 (1968).

15. G. V. Chester, M. E. Fisher, and N. D. Mermin, *Phys. Rev.* **185**, 760 (1969).

16. Y. Imry, *Ann. Phys. (N.Y.)* **51**, 1 (1969).

17. H. E. Stanley and T. A. Kaplan, *Phys. Rev. Lett.* **17**, 913 (1966).

18. D. Jasnow and M. E. Fisher, *Phys. Rev.* **B3**, 895, 907 (1971).

19. V. L. Ginzburg and L. P. Pitaevskii, *J. Exp. Theoret. Phys. (USSR)* **34**, 1240 (1958) [*English transl.: Sov. Phys.-JETP* **34**, 858 (1958)].

20. L. D. Landau, *J. Exp. Theoret. Phys. (USSR)* **11**, 592 (1941); L. D. Landau and E. M. Lifshitz, "Statistical Physics." Pergamon, Oxford, 1958.

21. B. D. Josephson, *Phys. Lett.* **21**, 608 (1966).

22. Yu. G. Mamaladze, *J. Exp. Theoret. Phys. (USSR)* **52**, 729 (1967) [*English transl.: Sov. Phys.-JETP* **25**, 479 (1967)].

23. E. P. Gross, *Ann. Phys.* **4**, 57 (1958); **9**, 292 (1960).

24. E. P. Gross, *J. Math. Phys.* **4**, 195 (1963).

25. G. Lasher, *Phys. Rev.* **172**, 224 (1968).

26. V. L. Berezinskii, *Sov. Phys.-JETP* **32**, 493 (1970); **34**, 610 (1971).

27. J. M. Kosterlitz and D. J. Thouless, *J. Phys. C.; Solid State Phys.* **5**, 124 (1972); **6**, 1181 (1973).

28. R. P. Feynman, *Progr. Low Temp. Phys.* **I**, (C. J. Gorter, ed.) p. 17. North-Holland Pub., Amsterdam 1955.

29. H. P. R. Frederikse, Ph.D. Thesis, Leyden, 1947.

30. H. P. R. Frederikse, *Physica* **15**, 860 (1949).

31. S. V. R. Mastrangelo and J. G. Aston, *J. Chem. Phys.* **9**, 1370 (1951).

32. D. F. Brewer, D. C. Champeney, and K. Mendelssohn, *Cryogenics* **1**, 1 (1960).

33. A. J. Symonds, Ph.D. Thesis, Sussex, 1965.

34. D. F. Brewer, *J. Low Temp. Phys.* **3**, 205 (1970).

35. M. Bretz, *Phys. Rev. Lett.* **31**, 1447 (1973).

36. S. Doniach, *Phys. Rev. Lett.* **31,** 1450 (1973).

37. O. Penrose, Onsager Festschrift, 1973.

38. E. A. Long and L. Meyer, *Phys. Rev.* **85,** 1030 (1952).

39. E. A. Long and L. Meyer, *Phys. Rev.* **98,** 1616 (1955).

40. D. F. Brewer and K. Mendelssohn, *Proc. Roy. Soc. (London)* **A260,** 1 (1961).

41. K. Fokkens, W. K. Taconis, and R. de Bruyn Ouboter, *Physica* **32,** 2129 (1966).

42. R. P. Henkel, G. Kukich and J. D. Reppy, *Proc. Int. Conf. Low Temp. Phys., 11th, St. Andrews,* 1968.

43. R. S. Kagiwada, J. C. Fraser, I. Rudnick, and D. Bergmann, *Phys. Rev. Lett.* **22,** 338 (1969).

44. M. Chester, L. C. Yang, and J. B. Stephens, *Phys. Rev. Lett.* **29,** 211 (1972).

45. J. A. Herb and J. G. Dash, *Phys. Rev. Lett.* **29,** 846 (1972).

46. C. H. Anderson and E. S. Sabisky, *Phys. Rev. Lett.* **30,** 1122 (1973).

47. M. Chester and L. C. Yang, *Phys. Rev. Lett.* **31,** 1377 (1973).

48. J. H. Scholtz, E. O. McLean, and I. Rudnick, *Phys. Rev. Lett.* **32,** 147 (1974).

49. M. H. W. Chan, A. W. Yanof, and J. D. Reppy, *Phys. Rev. Lett.* **32,** 1347 (1974).

50. K. R. Atkins, *Physica* **23,** 1143 (1957).

51. K. R. Atkins and I. Rudnick, *Progr. Low Temp. Phys.* **6,** 37 (1970).

52. C. W. F. Everitt, K. R. Atkins, and A. Denenstein, *Phys. Rev. Lett.* **8,** 161 (1962).

53. I. Rudnick, R. S. Kagiwada, J. C. Fraser, and E. Guyon, *Phys. Rev. Lett.* **20,** 430 (1968).

54. B. Ratnam and J. Mochel, *Phys. Rev. Lett.* **25,** 711 (1970); *J. Low Temp. Phys.* **3,** 239 (1970).

55. M. Revzen, *Phys. Rev. Lett.* **22,** 1102, 1413 (1969).

56. D. Bergmann, *Phys. Rev.* **A3,** 2058 (1971).

57. D. L. Goodstein and R. L. Elgin, *Phys. Rev. Lett.* **22,** 383 (1969).

58. J. G. Dash and J. A. Herb, *Phys. Rev.* **7,** 1427 (1973).

59. J. S. Langer and J. D. Reppy, *Progr. Low Temp. Phys.* **6,** 1 (1970).

60. H. E. Hall, *Phil. Trans.* **A250,** 359 (1957).

61. W. F. Vinen, *Proc. Roy. Soc. (London)* **A260,** 218 (1961).

62. P. J. Bendt, *Phys. Rev.* **127,** 4441 (1962).

63. J. D. Reppy and D. Depatie, *Phys. Rev. Lett.* **12,** 187 (1964).

64. G. Kukich, R. P. Henkel, and J. D. Reppy, *Phys. Rev. Lett.* **21,** 197 (1968).

65. M. W. Cole, J. G. Dash, and J. A. Herb, *Phys. Rev.* **B11,** 163 (1975).

66. J. A. Herb, Ph.D. Thesis, Univ. of Washington, 1974.

Author Index

Subject Index

A

Activity, absolute, 127
Adatom site lifetime, 20
Adatom states, 17–27
Adsorbate, definition, 60, *see also*
    individual gases
Adsorbents, *see* Substrates
Adsorption sites, 7
Alkali halide substrates, 54
    ethane adsorption, 124
    ethanol adsorption, 199
    helium adsorption, 26
    krypton adsorption, 124
    methane adsorption, 124
    water adsorption, 199
    xenon adsorption, 124
Argon adsorption
    copper substrates, 209
    glass substrates, 223
    graphite substrates, 124, 125, 217
    krypton substrates, 18
    titanium dioxide substrates, 46, 48,
        157
    tungsten substrates, 41, 43, 47
    zirconium substrates, 223

Argon solid substrate
    helium adsorption, 22, 26
    preplating, 48
Auger electron spectroscopy, 34
    coverage determination, 197

B

Band structure, adatom band states,
    19–27
Bandwith, ^{4}He on various substrates, 26
Beck potential, 140
Binding energy, 9–14, 41
Boron nitride substrates, 54
Bose–Einstein condensation
    ideal 2D gas, 102

2D gas in lateral fields, 228–232
Bromine on graphite, 195

C

Cadmium bromide, *see* Layer
    compounds
Calorimetry, 39, *see also* Heat capacity,
    Specific heat
Canonical ensemble, complete adsorp-
    tion system, 61